TRAVEL, TOURISM AND ART

Current Developments in the Geographies of Leisure and Tourism

Series Editors

Jan Mosedale, University of Sunderland, UK and **Caroline Scarles**, University of Surrey, UK in association with the Geographies of Leisure and Tourism Research Group of the Royal Geographical Society (with the Institute of British Geographers).

Tourism and leisure exist within an inherently dynamic, fluid and complex world and are therefore inherently interdisciplinary. Recognising the role of tourism and leisure in advancing debates within the social sciences, this book series, is open to contributions from cognate social science disciplines that inform geographical thought about tourism and leisure. Produced in association with the Geographies of Leisure and Tourism Research Group of the Royal Geographical Society (with the Institute of British Geographers), this series highlights and promotes cutting-edge developments and research in this field. Contributions are of a high international standard and provide theoretically-informed empirical content to facilitate the development of new research agendas in the field of tourism and leisure research. In general, the series seeks to promote academic contributions that advance contemporary debates that challenge and stimulate further discussion and research both within the fields of tourism and leisure and the wider realms of the social sciences.

Travel, Tourism and Art

Edited by

TIJANA RAKIĆ

Edinburgh Napier University, UK

and

JO-ANNE LESTER

University of Brighton, UK

Routledge
Taylor & Francis Group
LONDON AND NEW YORK

First published 2013 by Ashgate Publishing

2 Park Square, Milton Park, Abingdon, Oxon OX14 4RN
711 Third Avenue, New York, NY 10017, USA

Routledge is an imprint of the Taylor & Francis Group, an informa business

First issued in paperback 2016

British Library Cataloguing in Publication Data
A catalogue record for this book is available from the British Library

The Library of Congress has cataloged the printed edition as follows:
Rakic, Tijana.
Travel, tourism and art / by Tijana Rakic and Jo-Anne Lester.
pages cm. – (Current developments in the geographies of leisure and tourism)
Includes bibliographical references and index.
ISBN 978-1-4724-1039-9 (hardback)
1. Tourism and art–History. 2.
Grand tours (Education)–History. 3. Voyages and travels. I. Title.
N72.T68R35 2013
701'.08–dc23

2013019356

ISBN 978-1-4724-1039-9 (hbk)
ISBN 978-1-138-24633-1 (pbk)

Contents

List of Figures

Notes on Contributors

Jo Ankor, PhD has recently completed her doctoral research in Philosophy at the University of Technology, Sydney in which she applied philosophy to explore the phenomenon of tourism. Her research interests include tourism, creative practice and philosophy.

David Botterill, PhD is a Professor Emeritus at the Welsh Centre for Tourism Research, Cardiff Metropolitan University, UK and a Senior Research Fellow in the Oxford School of Hospitality Management, Oxford Brookes University, UK. He has published extensively in the areas of tourism and leisure studies.

Kevin Hannam, PhD is a Professor of Tourism Mobilities at Leeds Metropolitan University, UK and Visiting Senior Research Fellow at the University of Johannesburg, South Africa. He is founding co-editor of the journal *Mobilities* and, among other projects, he is also currently engaged with the Walking, Art, Landscape and Knowledge (WALK) project.

Brian Hudson, PhD is an Adjunct Professor of Urban Development at Queensland University of Technology, Australia. He has published widely on landscape and tourism, and is the author of *The Waterfalls of Jamaica: Sublime and Beautiful Objects* (2001) and *Waterfall: Nature and Culture* (2012).

Jo-Anne Lester, PhD is a Principal Lecturer at the University of Brighton. Her research interests include visuality and tourism, the relationships between travel and photography, visual methodologies and the consumption of tourist space with a particular interest in popular film as visual data. In addition to this volume, she has also co-edited *Mediating the Tourist Experience* (Ashgate, forthcoming, with C. Scarles) and two volumes on *Tourism and Visual Culture* (CABI, 2010, with P. Burns, C. Palmer and L. Bibblings).

Heather Mair, PhD is an Associate Professor in the Department of Recreation and Leisure Studies at the University of Waterloo, Canada. Her research interests include community development through tourism, leisure and sport.

John McCarthy, PhD is a Reader in Urban Studies at Herriott-Watt University, UK. He is an experienced spatial planner and an academic. He has conducted research focusing on culture entertainment and tourism-led regeneration, as well as authored a range of publications in these fields.

Pau Obrador, PhD is a Senior Lecturer in Tourism at the University of Sunderland. His research interests are in the theorisation of tourism, cultures of mass tourism and the role of the senses, mainly touch.

Marsha Pearce, PhD has recently completed her doctoral research in Cultural Studies at the University of the West Indies, Trinidad and Tobago. Her research interests include visual culture and issues of Caribbean representation and identity.

Tijana Rakić, PhD is a Lecturer at Edinburgh Napier University. Her research interests and publications predominantly lie in visual research methods, tourism and arts, history of tourism, and the relationships between world heritage, tourism and national identity. In addition to this volume, she has also co-edited *An Introduction to Visual Research Methods in Tourism* (Routledge, 2012, with D. Chambers) and *Narratives of Travel and Tourism* (Ashgate, 2012, with J. Tivers).

Mahsa Rouzrokh is a PhD candidate in the Department of Recreation and Leisure Studies at the University of Waterloo, Canada. Her research interests include the relationships between art, immigration and leisure.

Tony Seaton, PhD is a MacAnally Professor of Travel History and Tourism Behaviour at the University of Limerick, Ireland. He has published from, and taught widely at, four universities over the past 25 years: Northumbria, Strathclyde, Bedford and now Limerick. He holds degrees in Literature, Tourism Behaviour, Social Science and Theology.

Jean-François Staszak, PhD is a Professor of Geography at the University of Geneva, Switzerland. His earlier research lied within the history, epistemology of geography, and economic and cultural geography. His most recent work addresses geographical imaginaries in the fields of art and tourism, analysing the geographical othering process and the eroticization of the Exotic.

Elena Stylianou, PhD is a Lecturer in Art History and theory at the European University of Cyprus. Her research interests include contemporary art, visual culture, and museum studies.

Theopisti Stylianou-Lambert, PhD is a Lecturer in Multimedia and Graphic Arts at the Cyprus University of Technology and the coordinator of the 'Visual Sociology and Museum Studies Lab'. Her research interests include visual sociology, photography and museum studies.

Kristin M. Sullivan is a PhD candidate in Anthropology at the University of Maryland, USA. Her research focuses on the performance and presence of vernacular art as heritage in touristic settings.

Fiona Thompson is a ceramic Artist and a Writer. Her main focus lies in creating artwork for national and international exhibitions. She also undertakes commissions, residencies, teaching, collaborative projects and research.

Alexander Vari, PhD is an Assistant Professor of modern European History at Marywood University, USA. He has published a range of research articles in both peer-reviewed journals and edited volumes.

Stephen Williams, PhD is Emeritus Professor of Human Geography at Staffordshire University, UK. He is the author of a number of texts on tourism and recreation including *Outdoor Recreation and the Urban Environment* (Routledge, 1995), *Tourism and Recreation* (Prentice Hall, 2003) and *Tourism Geography: A New Synthesis* (Routledge, 2009).

Sharon Wilson is a Postgraduate Academic Assistant in Tourism undertaking a PhD on VW Campervan subcultures at the University of Sunderland, UK. She teaches in tourism and is also a visual arts practitioner and a museum consultant.

Acknowledgments

We would like to thank the Royal Geographical Society with the Institute of British Geographers and the Geographies of Leisure and Tourism Research Group for hosting a special session on Travel, Tourism and Art which we organised at the Annual RGS Conference in London in 2011 and where the idea for this book was born. We would also like to thank all the presenters at that session, as well as the numerous authors who responded to our subsequent open call for chapter proposals, many of whose proposals we were unable to accept due to the word count strictures of a single volume. We hope that the extraordinary interest in the broad theme of relationships between travel, tourism and art (which we witnessed during the various stages of planning, editing and finalising this book for submission to the publishers), will continue to grow and that this book will contribute to what we feel is a fascinating, and yet, still an underexplored theme of research in the social sciences and humanities.

Finally, we would also like to thank Yorgos Karagiannakis for helping us prepare the images contributors sent us for inclusion in their chapters according to publisher's guidelines, the book series editors for their valuable feedback during the initial stages of the book proposal, and the editorial and production team at Ashgate for their support throughout the various stages of our work on this book.

Chapter 1
Relationships between Travel, Tourism and Art: An Introduction

Tijana Rakić and Jo-Anne Lester

Art, in its many forms, has long played an important role in people's imagination, experience and remembrance of places, cultures and travels, as well as in their motivation to travel, while travel and tourism have also inspired numerous artists and featured in many artworks. Some of these themes have already received attention in academic research with studies focusing on different aspects of the relationships between travel, tourism and art published in a variety of formats including monographs (e.g. see Adams, 2006; Beeton, 2005; Tzanelli, 2007; 2013), journal articles (e.g. see Graburn, 1984; Cohen, 1993; Tribe, 2008; Obrador and Carter, 2010; Connell, 2012; Thompson, Hannam, and Petrie, 2012), edited volumes (e.g. see Graburn, 1976; Robinson and Andersen, 2002; Ruoff, 2006), book chapters (e.g. see art related contributions in Crouch and Lübbren, 2003; Crouch, Jackson, and Thompson, 2005; Tribe, 2009; Burns, Lester, and Bibbings, 2010; and Tivers and Rakić, 2012), and publications based on art exhibitions (e.g. see Dahlgren, Foreman, and Van Eck, 2005). While existing publications have included a consideration of a number of art forms, the ones which have seemingly received most attention to date are tourist arts (e.g. see Graburn, 1976, 1984; Cohen, 1993; and Adams, 2006), literature (e.g. see Robinson and Andersen, 2002) and film (e.g. see Ruoff, 2006; Beeton, 2010; Connell, 2012; Tzanelli, 2007, 2013).

Given the visual nature of many art forms, a number of existing studies explore representational aspects of the relationships between travel, tourism and art and the power of arts to reinforce and partly (re)construct particular images of a place as well as images of local or tourist cultures which these portray. Said's (1978) seminal work *Orientalism* and his polemic on the West's depictions of the Orient, for example, highlights the constituting power of historical travel accounts, literature and poetry in presenting and constructing the 'other'. In a similar vein, Thompson's (2006) study presents a detailed historical analysis of the ways in which art, such as postcards and photographs from the colonial period has both represented and 'tropicalized' Jamaica and the Bahamas, and by extension characterised the wider Caribbean. Contributing further to the theme of representation, although this time focusing on representations of tourism rather than locals, Tribe (2008) relies on the innovative methodological approach of 'virtual curating' and provides a rich insight into the representations of tourism in a range of artworks including paintings, film, photography, installation, performance, sculpture and graffiti, highlighting in such a way the importance of travel and tourism as a subject of particular interest to artists. Additionally, a book length publication from an art exhibition titled *Universal Experience: Art, Life and the Tourist's Eye* (Dahlgren et al., 2005) adds another layer of insight into the phenomenon of travel and tourism as an inspiration for artists, this time from the perspective of the art world. Interestingly, authors such as Thompson et al. (2012) discuss the production of a set of ceramic artworks in order to explore an unusual theme, that of the modes in which travel and tourism related research can act as an inspiration for artistic practice in which the artist seeks to include representations of travel and tourism.

Moving beyond the exclusively representational sphere, several insightful studies extend their consideration to include local artworks produced for tourism consumption (e.g. see the collection of essays in Graburn, 1976; contributions in Cohen's special issue, 1993; and Adams, 2006). Graburn's (1976) edited collection of essays explores arts and artisans of the Fourth World[1] and draws together a selection of cases which examine art practices and art products, often valued for their aesthetic qualities and frequently produced for tourists. Contributing further to this theme of enquiry, the special issue of the *Annals of Tourism Research* on tourist arts edited by Cohen (1993) includes contributions that bring to the fore a range of themes, such as commoditisation of culture, consumption practices, authenticity, symbolism and ideology, all of which highlight the significance of local art in the context of travel and tourism. More recently, this particular area of enquiry is further enriched by Adams's (2006) ethnographic study of Sa'dan Toraja (a minority group of predominantly Christian people living in Sulawesi, Indonesia) and their ethnic arts, which explores these art objects as symbols and markers of local identity.

Taking us into a different sphere of relationships between travel, tourism and art other studies include a consideration of public space artworks. For example, both Roberts (2012) and Andrews (2012) provide a rich insight into Antony Gormley's *Another Place* in their exploration of the liminality of beach space through an investigation of visitors' responses to and interactions with these 100 life-size cast iron figures which are a permanent art installation spanning 3km on Crosby beach, north of Liverpool, UK. Similarly, although in their case in the context of urban spaces, Lasansky and McLaren's edited volume (2004) reminds us that renowned architectural works can often become tourist attractions in their own right. In doing so they highlight not just the interplay between architecture and tourism but also '… how photography, film and souvenirs have been deployed to help mediate and mythologize specific sites – thereby blurring the boundaries between so-called "high" and "low" culture' (Lasansky, 2004, p. 3). Furthermore,

1 At the time, Graburn articulated the Fourth World as 'the collective name for all aboriginal or native peoples whose lands fall within the national boundaries and techno-bureaucratic administrations of the countries of the First, Second, and Third Worlds. As such, they are peoples who are usually in the minority and without power to direct the course of their collective lives' (1976, p. 1).

while remaining within the theme of urban spaces, tourism and art, Obrador and Carter (2010) emphasise the importance of artistic interventions in public spaces in the context of tourism through their discussion of the activities of a group of artists in Barcelona known as *Tactical Tourism.*

A final theme surrounding the relationships between travel, tourism and art from the existing literature which we will mention here is the one of the importance of artworks and artists in (at least partly) informing people's imagination and preconceived ideas about 'touristic' places, local and tourist cultures, and by extension also motivating people to travel to particular places. In this context of inducing travel, the art forms which have arguably attracted most scholarly attention to date are literature and film (e.g. see Squire, 1994; Herbert, 2001; Robinson and Andersen, 2002; Hudson and Ritchie, 2006; Beeton, 2010; Tzanelli, 2007, 2013; Carl, Kindon and Smith, 2007).

The fore-mentioned examples of existing studies and themes, albeit only briefly reflected upon here, serve to illustrate the multiplicity of relationships between different art forms, travel and tourism. Whilst acknowledging significant and insightful work in this area, we argue that what is absent within this vast and complex subject of inquiry is a critical mass of studies published within a single volume. This is the gap this edited volume aims to fill. It does so through: 1) an inclusion of a multi-(and at times inter-)disciplinary chapter contributions; 2) an inclusion of a range of relevant themes surrounding both historical and contemporary relationships between travel, tourism and art; 3) a recognition of importance of varying forms of art[2] in this context; and 4) an inclusion of studies relating to different geographical locations.

Our intention in editing a multi-disciplinary volume, despite the challenges this may present, is inspired by the fact that a topic as complex as this one, cannot possibly have a comfortable 'home' in only one academic discipline or field of study. Therefore, we believe that a deeper and richer insight into the relationships between travel, tourism and art can only be achieved through an inclusion of different disciplinary perspectives, or, as is the case with the inter-disciplinary studies included in this volume, through crossing of disciplinary boundaries. The academic disciplines and fields of study which have informed studies discussed in the various chapters in this volume include history, art history, anthropology, geography, philosophy, urban studies, tourism studies, cultural and leisure studies, while one of the chapters is co-authored by an artist and an academic (Chapter 12) and another is co-authored by two academics, one of whom is also a visual artist (Chapter 9).

2 The various forms of art discussed in this volume include both arts which would have traditionally been perceived as 'high' and 'low' (for a discussion on 'high' and 'low' arts in relation to tourism please see Wheeller, 2009). However, in wishing not to contribute to this division further, we have not labelled any of the art forms discussed in this volume as either 'high' or 'low'.

As mentioned earlier, this volume includes studies which focus on both historical and contemporary aspects of the relationships between travel, tourism and art. This is intentional in that we believe that the present (and potential future) relationships between travel, tourism and art can be better understood if underpinned by some knowledge of the history of arts in relation to travel and tourism and vice versa. While many of the chapters draw on, or refer to past events, four chapters focus on specific historical periods (Chapters 2, 3, 4, 6) and two focus on the importance of historical influence of arts and artists in contributing to the creation of particular destination images (Chapters 13 and 15). Finally, our intention to include studies that focus on a range of different art forms, rather than few selected ones, is also reflected in the actual chapters which include a consideration of graphic art (Chapter 2), literature (Chapter 3), poster art (Chapter 4), opera (Chapter 5), drawing (Chapter 6), carving (Chapter 7), public art (Chapters 8 and 15), *ad hoc* artistic installation (Chapter 9), painting (Chapters 10 and 13), photography (Chapter 11), ceramic art (Chapter 12), and film (Chapter 15).

Despite our and the contributors' attempts to offer as rich and diverse insights into a wide range of relationships between travel, tourism and art, many themes remain under-researched and some readers might also find inspiration for their own future studies on the pages of this book. With this in mind, this volume is organised in such a way so that it can be read cover-to-cover or selectively (depending on the particular topic(s) the reader is interested in). While there are inevitably overlaps between the different themes the chapters address, we have placed the chapters into four broad themes: i) representations of 'touristic' places, locals, travellers and tourists in artworks; ii) imagining, experiencing and remembering places through artwork; iii) travelling artists; and iv) art, place image and travel motivation. The broad themes and the chapter summaries included below will offer some guidance to those readers who wish to select particular topics of their interest, while the sequence of the chapters will be helpful to those readers who prefer reading this volume cover-to-cover. Whatever the chosen reading style, we hope that this volume and its chapters will be successful in being both a starting point and an inspiration for many scholars and students, as they embark on the fascinating journey of studying art in relation to travel and tourism or vice versa.

Part I, *Representations of 'Touristic' Places, Locals, Travellers and Tourists in Artworks*, opens with Tony Seaton's chapter titled *The Tourist Experience in Graphic Satire, 1796–1914* (Chapter 2). Seaton's chapter presents an analysis of graphic satire contained in collections of comic magazines and illustrated books produced and circulated during the late eighteenth- and early nineteenth- century Britain. His analysis of graphic art (supported by a consideration of archival material, secondary sources on graphic humour and catalogue raisonnés on caricature artists) traces representations of tourist behaviour in this particular historical period and explains the reasons for which graphic satire became the supreme medium which, in contrast to other art forms, allowed artists to represent psycho-social effects of tourism. Drawing on four theoretical psycho-social categories: sensations and perceptions; social interaction; self-presentation; and social class; Seaton supports his discussions with a range of illustrative examples and offers an insight into the modes in which aspects of tourist behaviour were depicted through this art form.

The following chapter titled *Lofty Hills Streaming With Water-Falls: The Tourist Landscape of Johnson and Boswell* and written by Brian Hudson (Chapter 3) is framed by the aesthetic concepts of the sublime, the beautiful and the picturesque. In it, Hudson explores the eighteenth-century writings of two literary artists, Samuel Johnson and James Boswell, as they toured Scotland together in 1773. Hudson presents an insightful analysis of Johnson's and Boswell's travel writing and reveals the importance of the then fashionable concepts of the sublime, the beautiful and the picturesque (which, as he reminds us, were also popular topics of intellectual discussions at the time), in their descriptive representations of waterfalls. In so doing, Hudson's chapter makes a contribution to our understanding of the role the concepts of sublime, picturesque and beautiful played not only in inspiring people to experience landscape features such as waterfalls, but also the role these concepts played in eighteenth-century British literary arts.

The *Art of the Railway Poster: Travel, Tourism and the Visual Representation of Place in Britain, 1920–1955* (Chapter 4) written by Stephen Williams geographically remains in Britain but focuses on a more recent historical period. Williams relies on a content analysis of approximately 400 railway posters published between 1920 and 1955 to demonstrate the importance of railway posters in shaping both the representation and the imagination of particular 'touristic' places and spaces by the travelling public, during a period which witnessed a significant growth in railway travel in the UK. He considers each poster in relation to five main variables, that of landscape or setting, presence or absence of people, characteristics of people, activity, and the moods each of the images sought to convey. Drawing on a range of examples, Williams provides an insight into the role railway poster art had in communicating particular representations of a number of places in Britain and, in so doing, highlights the power railway posters as a form of popular art has in reinforcing particular place identities and associated styles of tourism.

Chapter 5, titled *Towards a Realist Ontology of the Holiday: Seduction, Denial and Delusion in the Narratives of Tourism in Twentieth Century Opera* and written by David Botterill, concludes Part I of the volume. Botterill adopts the philosophical approach of scientific 'depth' realism and provides an ontological analysis of the holiday through an exploration of the representations of tourists in the narratives of a selection of twentieth century operas. He focuses on the characterisations of tourists in the opera librettos of The Death of Klinghoffer (Adams, 1991), Death in Venice (Britten, 1973), and Flight (Dove, 1998) and pays particular attention to arias which feature exchanges between the characters. This allows him to unearth situations which he names counter narratives of tourism, that of transgression, inversion and vulnerability in which the necessary conditions of a holiday, that seduction, denial and delusion are exposed. He argues that although necessary, seduction, denial and delusion are not sufficient conditions for a holiday to exist, and poses a series of questions regarding the reality of the holiday and the utopian ideal that is often associated with such pursuits.

Part II *Imagining, Experiencing and Remembering Places through Artwork* opens with a chapter written by Mahsa Rouzrokh and Heather Mair titled *Expeditions in Persia: Travel Journals and Artwork as Historical Documents* (Chapter 6). In their chapter, Rouzrokh and Mair discuss the significance of illustrative art, reproductions of which were often included in the eighteenth- and nineteenth-century travel journals, in enhancing representations and interpretations of the past, as well as, in motivating future travel to the places depicted. They present a study of Donald Wilber, his mid-twentieth century visit to Iran, and subsequent 1962 publication titled *Persian Gardens and Garden Pavilions*. They explain that while Wilber's expedition and publication were inspired by earlier travellers and their artwork, he also relied on his own experiences and a variety of other sources. These included other traveller accounts; eighteenth- and nineteenth-century artwork; miniature paintings from as far back as the fifteenth century; and even poetry. Such sources, as Rozrourk and Mair argue, all collectively contribute to the constructed knowledge of the past.

The next chapter in Part II is Chapter 7 titled *Carving the Past: Art, Tourism and Heritage in Chincoteague, Virginia* in which Kristin M. Sullivan delves into the continuing tradition of production of hunting decoy and decorative wildfowl carvings in Chincoteague, USA. She discusses the existence of these decoys and carvings not just as objects of art which are sold to tourists, but also as objects of particular interest to anthropologists due to their role in preserving and maintaining particular aspects of both local identity and heritage. In her study, Sullivan moves beyond the early anthropological literature on arts and tourism which downplayed the agency and power of both the artists and the local community in influencing the production of local artworks for tourism consumption. Instead, she suggests that those members of the local population involved in the process are actively engaged in constructing their identity and negotiating the values of their heritage through the production and distribution of hunting decoy and decorative wildfowl carvings.

The last chapter in Part II is titled *Welcome to Wherever: Art in an Airport* (Chapter 8) and is written by Jo Ankor. Taking the example of Adelaide Airport terminal, South Australia, Ankor discusses the installation of representational art panels as markers of cultural and social identity and the consumption of such public works of art in the liminal space of the airport terminal, a space that is both a gateway to, and exit from the destination. In her discussions she explores the challenges inherent in the placement of art installations in such spaces, and the potential issues which may arise. Through a theorisation of the nature of public and liminal spaces such as airports in general and Adelaide airport in particular, her chapter explores the role of, and potential controversies surrounding the power of artworks to infuse meanings of place within an airport space.

Part III *Travelling Artists* commences with Sharon Wilson's and Pau Obrador's chapter titled *The Nomadic Village: Communal Creativity and Political Subversion in a Temporary Settlement* (Chapter 9). Rooted in both theoretical discussions and findings drawn from a 'nomadic' ethnographic fieldwork their chapter discusses the *Nomadic Village* project during which 30 artists from a number of different countries came to stay and work in a mobile living space set up within the area of the village of Wolsingham, County Durham, UK. Given the nature of this project and the empirical approach Wilson and Obrador chose to take (the lead author was also a visual artist herself and therefore able to integrate and participate in nomadic artistic activities during the fieldwork stage of their research project) added further richness to the empirical data discussed in their chapter. Wilson's and Pau's discussions revolve around the notions of creative nomadism, the relationships between travel and creativity and the political, and at times paradoxical aspects of contemporary forms of creative nomadism, illuminating in such a way nomadic artistic practices and their relationships with the creative and transformative aspects of travel and tourism.

The following chapter, titled *Tropical Impressions of a Traveller: Caribbean Place and the Retouching of Sensory Experience in the Paintings of Steve Bonner* (Chapter 10), is written by Marsha Pearce. Her chapter theorises the concept and practice of retouching in the context of Steve Bonner's paintings and representations of the Caribbean. Relying on a variety of sources, such as content on Bonner's website, interviews with the artist himself, and an examination of his artwork, Pearce places a particular emphasis on the philosophical notion of reality as being relative and informed by subjective experiences. It is in this context that she places and interprets Bonner's travel to, and artwork about the Caribbean, claiming that his works, which are primarily produced for the holidaymakers looking for visual reminders of their stay in the Caribbean, are nonetheless phenomenological expressions of his own subjective experiences of the Caribbean place. The approach, which enables Bonner to produce artwork richly infused with his own subjective impressions of the Caribbean is, as Pearce demonstrates in this chapter, his practice of retouching.

Chapter 11, written by Theopisti Stylianou-Lambert and Elena Stylianou moves on to the discussion of Martin Parr's collection of tourism related photographs produced as a part of his artistic project and book publication *Small World* (2007[1995]). Titled *Martin Parr: A Traveller-Critic and a Professional Post-Tourist in a Small World,* their chapter relies on both modern and postmodern theories of tourism to offer a new understanding of Parr's role as an artist whose work revolves around tourists and touristic practices. Their analysis relies on a consideration of secondary sources which discuss Parr's work, a semiotic analysis of images included in Parr's 2007 edition of the book *Small World*, and Parr's own commentary about his photographic practice and offers new insights into the identity of Martin Parr as a professional photographer, a traveller-critic, and a professional post-tourist. His artwork, as Stylianou-Lambert and Stylianou demonstrate, moves beyond the modernist understandings of tourists as passive consumers and encompasses postmodern understandings in which tourists are perceived as self-aware producers of 'a polysemic tourism stage'.

The following chapter, which is also the last chapter in Part III is Chapter 12 titled *Hunting the Touristic Animal through Ceramic Art in Alberta, Canada* and is written by Fiona Thompson, a ceramic artist and Kevin Hannam, an academic. Through a case study of Thompson's own ceramic artwork, this chapter explores the connections between the artist's travel experiences to Alberta, Canada, tourism theories, and the practice of producing taxidermic ceramic plates. What is particularly interesting in Thompson and Hannam's chapter is the fact that the artwork discussed is inspired both by academic tourism research, as well as the artist's own visits/field trips to Alberta, Canada. In discussing existing academic literature on hunting and tourism, the significance of field trips and the various aspects of the artistic ceramic production process, Thompson and Hannam also highlight the appropriateness of the ceramic art form in the context of commemorating both animals themselves and the artist's own experiences of visiting Alberta.

The last part of this volume, Part IV *Art, Place Image and Travel Motivation* commences with Jean-François Staszak's chapter titled *The Artist and the Tourist: Gauguin in Tahiti* (Chapter 13). With its focus on exploring the role of art in informing geographical imaginings, this chapter presents a critical foray into Gauguin's life and work, his representations of Tahiti and the discursive power of his paintings in the continued exoticising of Tahiti as a tourist destination. Relying on a study of Gauguin's biography through previous publications about his life and art, an examination of a selection of his paintings and his own experiences from several visits to Tahiti, Staszak discovers striking similarities between Gauguin's imaginings of Tahiti and that of the geographical imaginaries typical of his time. Contrary to popular belief which often perceives tourists as following the paths of artists as far as the imagination of 'touristic' places is concerned, in his chapter Staszak suggests that, at least in the case of Gauguin and Tahiti, artists' and tourists' imaginations of place may have much in common.

The second chapter in this part of the volume, *Image, Identity and Tourism: Public Art in Cultural Quarters* (Chapter 14) is written by John McCarthy. His chapter takes on an interesting topic of the role of public art in creating 'cultural quarters' and achieving broad urban regeneration initiatives given that public art is often perceived to be successful in attracting visitors, enhancing local identities, as well as promoting a particular place image. He relies on two UK case studies, Manchester's cultural quarter called 'Northern Quarter' and Belfast's 'Cathedral Quarter' in order to explore the effectiveness of public art in this context, as well as, pose a question of whether aims of creating a particular image of a quarter can be contradictory with its local identity. Whilst challenging aspects of 'culture-led' approaches through cultural quarters, McCarthy also claims that public art can be effective in contributing to 'culture-led regeneration', tourism development, and, in enhancing the overall quality of life. McCarthy emphasises both the inherent challenges as well as offers potential solutions which can contribute to the success of such endeavours on economic, social and environmental levels.

The chapter which closes the fourth theme and the volume as a whole is Chapter 15 titled *Branding Montmartre: Tourism, Popular Culture, and Artistic Bohemia*. Written by Alexander Vari, this chapter contains an insight into the long-term process of branding Montmartre, a Parisian neighbourhood which was home to its best known artistic cabarets including the *Cabaret du Chat Noir* and the *Moulin Rouge*, the birthplace of the modern avant-garde art and a neighbourhood where artists such as Van Gogh, Toulouse-Lautrec, Picasso and Modigliani once lived and painted. Since then, Montmartre has become both a Parisian and global cultural icon. Through a reliance on a wide range of historical and contemporary sources and references to a number of films, Vari demonstrates that the long-term process of branding Montmartre as a bohemian neighbourhood best known for its turn-of-the-century cultural developments was aided by numerous cultural actors. These, according to Vari, included the artists themselves and later also extended to include participants such as film makers, memoirists, and media conglomerates, all of whom played an important role in keeping the myth of *fin-de-siècle* Montmartre alive and in creating a particular brand which attracts millions of tourists and ignores the present-day realities of the neighbourhood.

References

Adams, K. A. (2006). *Art as Politics: Re-crafting Identities, Tourism, and Power in Tana Toraja, Indonesia*. USA: University of Hawai'i Press.

Andrews, H. (2012). Another Place or Just Another Space? Liminality and Crosby Beach, in H. Andrews and L. Roberts (eds). *Liminal Landscapes: Travel Experience and Spaces In-between*, (152–166). London: Routledge.

Beeton, S. (2005). *Film-Induced Tourism*. Clevedon: Channel View Publications.

Beeton, S. (2010). Introduction: The Advance of Film Tourism. *Tourism Planning and Development*, 7(1), 1–6.

Burns, P., Lester, J., and Bibbings, L. (eds) (2010). *Tourism and Visual Culture: Volume 2 Methods and Cases*. Wallingford: CABI.

Carl, D., Kindon, S., and Smith, K. (2007). Tourist's Experiences of Film Locations. *Tourism Geographies*, 9(1), 49–63.

Cohen, E. (1993). Introduction: Investigating Tourist Arts. *Annals of Tourism Research*, 20(1), 1–8.

Connell, J. (2012). Film Tourism – Evolution, Progress and Prospects. *Tourism Management*, 33(5), 1007–1029.

Crouch, D., Jackson, R., and Thompson, F. (eds) (2005). *The Media and the Tourist Imagination: Converging Cultures*. London: Routledge.

Crouch, D. and Lübbren, N. (eds) (2003). *Visual Culture and Tourism*. Oxford: Berg.

Dahlgren, K., Foreman, K., and Van Eck, T. (2005). *Universal Experience: Art, Life and the Tourist's Eye*. Chicago and New York: Museum of Contemporary Art and Ditributed Art Publishers.

Graburn, N. H. H. (1984). The Evolution of Tourist Arts. *Annals of Tourism Research*, 11(3), 393–419.

Graburn, N. H. H. (ed.) (1976). *Ethnic and Tourist Arts: Cultural Expressions from the Fourth World*. Berkeley: University of California Press.

Herbert, D. (2001). Literary Places, Tourism and the Heritage Experience. *Annals of Tourism Research*, 28(2), 312–333.

Hudson, S. and Ritchie, J. (2006). Promoting Destinations Via Film Tourism: An Empirical Identification of Supporting Marketing Initiatives. *Journal of Travel Research*, 44, 387–398.

Lasansky, M. (2004). Introduction, in M. Lasansky and B. McLaren (eds). *Architecture and Tourism: Perception, Performance and Place*. Oxford: Berg.

Lasansky, M. and McLaren, B. (eds) (2004). *Architecture and Tourism: Perception, Performance and Place*. Oxford: Berg.

Obrador, P. and Carter, S. (2010). Art, Politics, Memory: Tactical Tourism and the Route of Anarchism in Barcelona. *Cultural Geographies*, 17(4), 525–531.

Roberts, L. (2012). Another Space: Gleaning the Urban Littoral. *Journal of Tourism Consumption and Practice*, 4(1), 94–100.

Robinson, M. and Andersen, H. C. (eds) (2002). *Literature and Tourism: Essays in the Reading and Writing of Tourism*. London: Thomson.

Ruoff, J. (ed.) (2006). *Virtual Voyages: Cinema and Travel*. Durham and London: Duke University Press.

Said, E. W. (1978). *Orientalism*. London: Penguin.

Squire, S. J. (1994). The Cultural Values of Literary Tourism. *Annals of Tourism Research*, 21(1), 103–120.

Thompson, K. A. (2006). *An Eye for the Tropics: Tourism, Photography, and Framing the Caribbean Picturesque*. Durham and London: Duke University Press.

Thompson, F., Hannam, K., and Petrie, K. (2012). Producing Ceramic Art through Tourism Research. *Annals of Tourism Research*, 39(1), 336–360.

Tivers, J. and Rakić, T. (eds). (2012). *Narratives of Travel and Tourism*. Farnham: Ashgate.

Tribe, J. (2008). The Art of Tourism. *Annals of Tourism Research*, 35(4), 924–944.

Tribe, J. (ed.) (2009). *Philosophical Issues in Tourism*. Bristol: Channel View Publications.

Tzanelli, R. (2007). *The Cinematic Tourist: Explorations in Globalisation, Culture and Resistance*. New York: Routledge.

Tzanelli, R. (2013). *Heritage in the Digital Era: Cinematic Tourism and the Activist Cause*. New York: Routledge.

Wheeller, B. (2009). Tourism and Arts, in J. Tribe (ed.). *Philosophical Issues in Tourism* (pp. 191–208). Bristol: Channel View Publications.

PART I
Representations of 'Touristic' Places, Locals, Travellers and Tourists in Artworks

Chapter 2
The Tourist Experience in Graphic Satire, 1796–1914

Tony Seaton

Introduction

This chapter traces how and why, during the long nineteenth century, graphic satire became the supreme visual medium for representing the psycho-social effects of tourist experiences in ways that were neither attempted, nor possible, in any other art form. Tourism was a radical, new mobility system that emerged from technological, social and ideological developments in the late eighteenth and early nineteenth centuries (Towner, 1996). It was quickly espoused by an expanding, English, middle class created in the first great century of capitalism, and later spread down the social scale (Bailey, 1978; Cunningham, 1980). The tourist role did not just affect the physical movement of people, but their psychological orientation and social relations. In much the same way, as Simmel (1950) has argued, that city living modified the consciousness and mental orientation of urban dwellers, so the new, travelling protocols of tourism produced new modes of perception, sensation, social interaction, and self-presentation.

Simultaneously, the revolution in behaviour and consciousness brought about by modernising forms of travel was accompanied by another revolution: the growth of mass communication through the production and commercial dissemination of cheap printing and graphic art, which included cartoons and comic illustration (James, 1976). The chapter describes and exemplifies the synergy between these two developments that resulted in an immense output of graphic humour, documenting the psycho-social effects of tourism during the century that shaped its modern form, and which now constitutes a matchless, and largely under-valued, resource for the social historian.

The chapter derives its main arguments and its illustrative examples from a qualitative appraisal of archival material, reference works on graphic humour, catalogue *raisonnés* on caricature artists, and collections of comic magazines and illustrated books in Britain.

Travel, Tourism and Art

Until the rise of the industrial middle classes in the late eighteenth and nineteenth centuries, tourist experiences rarely figured in art. The daily activities of the aristocratic, Grand Tourists who had dominated travel previously left few traces, in contrast to the destinations they visited, particularly in Italy, France and Switzerland, which were very often painted and reproduced through engravings. The main visual record of Grand Tourists was typically made when they returned home and posed for portraits in front of neo-classical, country houses, or grouped with fellow dilettantes at library tables, displaying objects connoting antiquarian travel and/or scientific interests.[1] Neither revealed much about their touring experiences, the focus being on the consumer effects: the clothes, architectural styles and interior designs for home and garden improvements, acquired while they were away.

One reason for the limited representation of the tourist in art was that paintings were mainly sponsored by social elites as public displays of conspicuous taste that did not include recording the minutiae of everyday living, whether at home or touring abroad. The preferred subjects of paintings were ones that conformed to established genres: the religious narrative; the portrait; the conversation piece; the classical landscape; the history painting. Such works signified the correct tastes of the patron along with their status, public interests and lifestyles, including their ownership of commodities, property, animals and women. Painting on canvas was the artistic 'gold standard' for achieving these celebrations of the aristocratic and gentrified self, quintessentialising them in images that hung as frozen assets on the wall.

The other factor that kept the day-to-day experiences of the tourist out of the painter's frame was a practical one, relating to the inherent limitations of the medium. Painting offered relatively static images that could not represent the movement and changing scenes that constituted the emerging mobilities of tourist behaviour. Painting required careful preparation of canvas and materials, preliminary designs and heuristic experiments that might involve false starts and adjustments, before the progression of a work that might take months, or even years, to complete. Necessity, as much as ideology, linked painting with the essentialisms of *static* subjects, rather than representations of transitory, more mobile ones. The painter's role was accordingly celebrated as that of quasi-visionary, equipped to disclose *one* immanent and timeless truth about their subject, whether: revealing the character, social importance or transcendent beauty of the person in a portrait; evoking the mythologised spirit of place and time in a classical landscape; or capturing the decisive moment in a history painting. Such miracles of teleological summation were hard to distinguish in the rapidly changing mobilities of tourism. Tourism and tourists resisted static, quintessentialising summary.

1 In 1742 a group of young aristocrats and gentry established the Society of Dilettantes, a club for Grand Tourists like themselves who had 'seen Italy'. In 1914 a history of the Society included more than a dozen portraits of its members, reproduced from original paintings in museums and country house collections. Not one of the members was depicted touring. Most were single portraits, except for two group pictures by Sir Joshua Reynolds of dilettantes gathered around a table, wine glasses in hand, among instruments and antiquarian objects of virtu.

Graphics and the New Mobilities of Tourism

Tourism was a new system of mobilities that transformed the phenomenal world of its subjects physically, socially and mentally. It was rapid movement in space and time requiring repertoires of improvised behaviour in fleeting encounters with people, things and events. Tourists experienced the unpredictability of travel such as delays, poor roads, storms at sea, mishaps, queuing, overcrowding, as well as frustrating exchanges with service functionaries such as waiters, frontier guards, customs officials and hoteliers. They also had to manage brief encounters with fellow tourists with whom they might be thrown together. The tourist role thus demanded many competences and decisions, not least an increasing range of consumer choices before and during the tour. They included knowing what to wear and take on the journey, and how to cope with physical logistics that might include finding rooms, bargaining for seats, and successive transfers between different forms of transport. In addition, tourists sought ideological help on what tourism was ultimately about, which is why there were texts that taught the newly, enfranchised traveller how to travel, what to see and look for, and what to avoid (Kitchiner, 1827; Eldon, 1828; Fosbroke, 1826).

These new behavioural demands had wide-ranging, psycho-social consequences, that were similar to, and greater than, those which Schivelbusch attributed to the psychological impact of the railway (itself one of tourism's new mobilities) when he observed:

> … in little over a generation … (it) had introduced a new system of behaviour: not only of travel and communication but of thought, of feeling, of expectation. (Schivelbusch, 1980, p. xiii)

Graphic Humour and the Tourist

These new mobilities, imposed by tourism, were ones that painting, with its laborious, production processes and static effects, could hardly begin to represent. But there was a subaltern artistic form that could. This was graphic satire, a generic commodity that had emerged just after the invention of printing, and had first appeared in books, pamphlets and broadsides, as part of the propaganda wars, fought between Catholics and Protestants in the sixteenth century, and during the Civil War in England (Klingender, 1943; Grand-Carteret, 1927–1928). By the end of the eighteenth-century graphic satire had expanded as popular entertainment that satirised, not just religious and political events, but all kinds of fashionable fad and folly (George, 1967). One of its most popular subjects was the burgeoning and still, to many, partly mystifying practice of tourism, that being the fashion for going places you had no need to visit.

Graphic humour could depict and disseminate changing movements in time and space more effectively than higher art forms through its characteristics of *rubricity* and *multiplicity*. *Rubricity* was the way in which graphic humour could support one or more visual images with *words* by captions, commentary, or subjoined dialogue. Though paintings occasionally included verbal texts, this was neither normative nor common. In fact, in graphic humour this was both commonplace and expected. This word-and-picture facility routinely gave cartoons a more explicit and explanatory potential that reduced the polysemy which potentially exists in all visual images, unsupported by verbal augmentation. And it was able to do so in presenting subjects and narratives involving movement in time and/or space. A second and related attribute of graphic humour, unavailable to painting, was its *multiplicity*, being able to offer on a single page, or over a number of sequential ones, *several images* that could denote different, physical positions and spatial locations ordered in time. This was ideally suited to represent the mobilities of tourism. Figure 2.1 shows the combination of rubricity and multiplicity that could uniquely represent spatial and temporal change. It depicts how a restless night in a train, racing north to the Lake District, could be evoked by six different images that signified the shifting positions of disturbed passengers at different moments through the night (Atkinson, 1878).

The power of word-and-picture sequences to convey movement in time and space, prefigured the development of, what was to become, the strip cartoon, the apotheosis of 'graphic art's multiplicity' and rubricity (Kunzle, 1968, 1994). Both features gave graphic humour its dynamic capacity to capture the quotidian, world-

Figure 2.1 Incidents of the night journey

Source: Atkinson, 1878, plate 3.

in-motion of the tourist, with its quick changes of scene and transitory encounters. The result was the production of a growing torrent of humorous, illustrated reportage narrating, the carnivalesque (see Bakhtin, 1984) mobilities of tourist behaviour; the trouble and tribulations of ascending a mountain; developing games and intimacies on a beach; the worsening stages of sea sickness; the burlesqued difficulties of putting a tent up in a high wind; and many other tourist experiences requiring dynamic, progressive narration.

Graphic humour did not offer the same complexity as painting, but it did not need to. Caricature, as Gombrich (1960) was the first to recognise, did not depend on highly finished mimesis, drawn from close observation and *copying* of 'nature', but *translations* into invented, visual codes and original schema, pared down in an 'abbreviatory style', the purpose of which was the single minded communication of one, or a few, main 'points' about a person or narrative. If the 'comic artist's inventions were successful, the public quickly learned to play the game of decoding and enjoying the visual shorthand of graphic humour. As Gombrich commented:

> … [The] artist who uses an abbreviatory style can always rely on the beholder to supplement what he omits …[There is] a willingness of the public to accept the grotesque and simplified partly because its lack of elaboration guarantees… the absence of contradictory clues. (1960, p. 229 and 336)

What graphic cartoons lacked in 'finish' they, therefore, made up for in instant impact and discursive clarity, and the speed with which they could be delivered, through printing processes, to large audiences. The artist rapidly sketched his (in this historical period most graphic artists were men) comic inventions, which were then transferred by engraving on wood, steel or copper, or by aquatint and lithography, to paper. Copies could then be printed off within hours and sold in quantity as duplicated images, at a fraction of the cost of paintings. Graphic satire was thus a demotic force that reached wider audiences than 'high' art, its outputs circulating as commodities within a market system dependent, not on rich sponsors, but paying customers. This meant that its content had increasingly to reflect, or address, the tastes, interests and experiences of its audiences, rather than serve the interests of elite patrons.

The Media of Graphic Satire

During the 'long' nineteenth century with which this chapter is concerned, the graphic humour of tourism described above, circulated in 4 forms: as single prints, in comic journals, as illustrations in book length tour accounts, and as comic postcards.

The Single Print

Comic prints, sold individually or in sets, had flourished from the days of William Hogarth, the first major artist to make caricature a highly commercial genre

(Paulson, 1965; Atherton, 1974; Uglow, 1997). These were sold by booksellers, who exhibited them in shop windows where they regularly attracted large crowds. Many ended up in scrap books or framed on walls. In the last years of the eighteenth century, and the first two decades of the nineteenth, three great and prolific, English caricaturists were all publishing individual prints and sets that featured the emerging practices of tourism. They were James Gillray, Thomas Rowlandson and George Cruikshank (Everitt, 1886).

Gillray was known mainly for his savage, political cartoons (George, 1959; Atherton, 1974), but his work included a small number of memorable tourism representations, *Morning promenade upon the cliff at Brighton*, *Posting in Scotland* (1805) and *Posting in Ireland* (1805) (Wright and Evans, 1851).

Thomas Rowlandson produced a large volume of topographical engravings and drawings, individually and in sets, as well as illustrations for books, that featured social life at tourist destinations. They can be readily sampled by skimming the pages of the two volume catalogue of his work by Joseph Grego (Grego, 1880). It includes quintessential works such as *The Beauties of Boswell* (1786), a burlesque satire on the tour to the Hebrides made by Dr. Johnson and James Boswell, and *The comforts of Bath* (1798). His artistic output on travel, topography and the social scene from the 1780s until his death in 1827 was so great that his admirers speak of 'Rowlandson's England' in the same way that Jane Austen's fans speak of 'The England of Jane Austen'.

The work of George Cruikshank was focused most on the many peopled, social life of London, including his illustrations of *Oliver Twist* and *Sketches by Boz*, but it also included many sketches of fashionable resorts, burlesqued holiday incidents and illustrations for a celebrated, comic book on the Great Exhibition which was the greatest, international tourism occasion of the 1850s (Mayhew and Cruikshank, 1851).

The Comic Tour

The second form of circulation was the comic tour, a book length text, accompanied by illustrations. Hogarth's *Peregrination*, an account of a five day jaunt with four friends through Kent in May 1732, was the first example, originally intended as a private record to which each member of the party contributed words or pictures. It was not engraved and published until 50 years later (Hogarth, 1952).

The first important illustrated comic tour produced for public consumption was, *Eccentric Excursions* by George Mouthard Woodward (Woodward, 1796). This is a landmark work for the social historian, particularly the tourism historian, which has not been accorded the importance it deserves. It is a travelogue illustrated by 100 comic, full page plates, of a tour through different parts of England and South Wales, divided into 22 chapters by county. In it Woodward set out to mock the growing fashion for searching out picturesque scenery by choosing, instead, to focus on people, the tourists themselves, and the people they met on their travels. He put it this way:

> The public have been highly gratified by the works of an able and elegant author descriptive of the picturesque beauties of *landscape* and *cattle* illustrated by suitable designs: – then why should not the variegated forms of Nature, in the *physiognomy* of the human race in respect to *countenance* and *character,* afford subjects equally worthy of the *pen* and *pencil*. (Woodward, 1796, p. ii)

The 'able and elegant author' was William Gilpin, whose books on picturesque landscape had been instrumental in popularizing the vogue for touring by amateur painters searching for picturesque scenery. Woodward's aim was not just to satirise picturesque fashions. He hoped to capitalise on the vogue for physiognomy, the bogus science of reading character from facial features that had been created by the publications of J. C. Lavater; publications which achieved widespread influence in France, Germany and England between 1780s and 1830s. Woodward asserted that tourists needed to read the character of the people they met on their travels, and supplied his readers with burlesque cameos of odd people and burlesque incidents, encountered on tour through the English counties. The book was a great success and was reprinted nine times between 1796 and 1819. Today it offers matchless insights into the sensations, sights, sounds and conditions of travel in England and South Wales at the end of the eighteenth century.

After 20 years of success *Eccentric Excursions* was largely forgotten. But it was followed by a three volume set of illustrated, comic tours that achieved both immediate and lasting success, due to the genius of Thomas Rowlandson. *Doctor Syntax's Tours* featured the adventures of an impoverished vicar who set out on an artistic tour of England which was to begin in the Lake District, in order to raise money when he got home by publishing the sketches he had made en route (See Figure 2.2). Like Woodward's *Eccentric Excursions*, its aim was to make fun of the fashion for publishing picturesque tour memoirs and view books. Rowlandson's engraved sketches of the mishaps and misfortunes that befell the hopeful vicar were accompanied by a text in verse by a hack journalist called Combe. The works were immensely popular and were reprinted throughout the century and spawned many imitations set in France, Scotland and London.

The greatest, formal innovation in illustrated, comic tours was made not in England, but in Switzerland, by an artist named Rudolphe Topffer. In the 1830s and 1840s he pioneered the strip cartoon, an episodic narrative told entirely in pictures with brief, subjoined captions. He published a series of cartoon tours each of which featured the travelling quests and picaresque adventures of single, eponymous, male anti-heroes with ludicrous names that included: Monsieur Vieux Bois, Monsieur Pencil, Monsieur Jobard, Monsieur Crepin, and Monsieur Cryptogramme.

Topffer's work was quickly plagiarised. *Monsieur Vieux Bois* was translated as *Mr Oldbuck*, re-engraved, and issued anonymously by the publisher, David Bogue, in the early 1840s. More significantly, in 1854 Richard Doyle, who had made his name as a teenage prodigy on *Punch*, adopted the strip cartoon in an original work called, the *Tour of Brown, Jones and Robinson* (Doyle, 1854). This tale of three

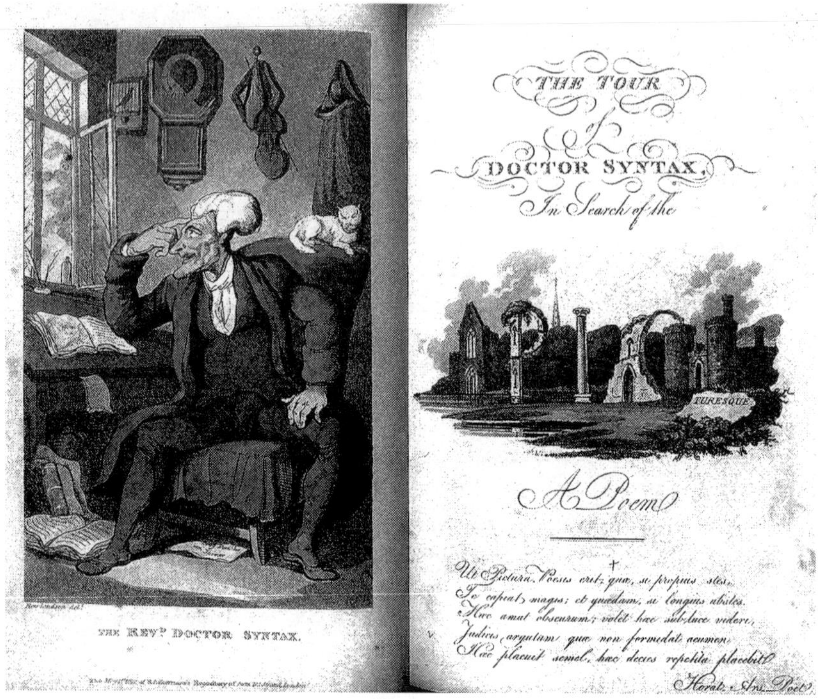

Figure 2.2 Dr Syntax

Source: Rowlandson, 1812, Volume 1, title page.

male friends on the loose in Europe, getting into one scrape after another, won great popularity, and the strip cartoon format became a widely used one in presenting comic tours until the end of the nineteenth century. Though the strip cartoon format was mainly adopted by professional cartoonists, it was also copied by amateurs, who kept illustrated journals of their tours with family and friends, and occasionally published them in small editions, intended mainly for private audiences.

The Comic Magazine

Comic magazines were the single biggest source of graphic humour during the reign of Queen Victoria. They had first appeared in the late 1830s, but most were short-lived and it was not until the 1840s and 1850s that they took hold and eventually became something of a national institution. The breakthrough was made by *Punch* magazine which was launched in 1841. It combined short, pithy articles on topical events, fashions and politics with full page cartoons (including a main political one), and smaller ones that often featured dialogue. *Punch* had many unsuccessful competitors but a number of them, including *Judy,* and *Fun* did survive, and from the 1850s comic magazines became a permanent feature of newsagents, including crucially, the railway book stands that were the conduit for a new kind of literature, railway reading.

Tourism became a mainstay of all these journals, particularly from June through September, when holiday cartoons acted as a carnivalesque marker that summer had arrived. People grew so used to holiday jokes in comic journals in summer time that a cartoon in *Judy* in 1869 showed a young woman surreptitiously posing for an artist she believed (wrongly) to be employed by a comic journal. By the end of the century *Punch* had published so many cartoons in its weekly pages that it was able to re-issue them in themed, book length compilations by subject. They included several tourism related subjects, among them *Punch on Holiday, Punch at the Seaside* and *Punch in Scotland.*

The Comic Postcard

The last great innovation in the graphic history of tourism was the comic postcard, which erupted symbolically in 1901, the year that Queen Victoria died. Comic postcards were the first graphic products to be consumed mainly by the working classes, and they achieved a vulgarity of content, now nostalgically remembered as 'saucy' or 'naughty' (Wykes, 1977; Barker, 1988). They depicted a burlesque, sexualised world, peopled by monstrous bodies of overweight ladies with shrivelled, hen-pecked husbands, loutish young men ogling busty young women, and men and women of all ages making suggestive double entendres on the beach, in hotel bedrooms and seaside toilets. Sending home comic postcards became a holiday ritual for workers until late into the twentieth century. The work of the most prolific postcard artist, Donald McGill, attracted the attention of George Orwell in 1941. In his essay-length study Orwell placed the comic postcard within a distinctive, proletarian way-of-life, and saw its vulgarity as an expression of the temporary, home-grown freedoms of English workers at play, out of reach of middle class norms and restraints. The argument anticipates the concept of resistance through class-based rituals, developed in the theorising of Cultural Studies in the 1970s. Other studies of comic postcards have followed (Alderson, 1970; Calder-Marshal, 1966; Green, 1976; Staff, 1979), as well as monographs on McGill's work (Buckland, 1976, 1984). One aspect of postcards that has so far escaped study is the contrast between the main printed text and the handwritten messages on the reverse. The anarchic vulgarity of the former is often at odds with the relatively prim conventionality of the 'Wish you were here/Weather bad' message of the latter.

Categories of Psycho-social Effect

The images represented in these four graphic genres offer a vast, textual field of exploration for researchers interested in the history, and psycho-social effects, of tourism during its modern growth period and the remainder of this chapter appraises some of the ways in which they do so.

Methodologically, the appraisal began by the selection of four, theoretical categories identified by psychologists and sociologists as important ones in understanding social behaviour: sensations and perceptions; social interaction; self-presentation and social class. The appraisal qualitatively explored how selected tourism images from the four genres of graphic humour, published between 1796 and 1914, represent the behavioural world of the tourist within each of the psycho-social categories. These chosen categories are just four of many conceptual categories which social scientists have deemed central ones in understanding social life. Both the ideas and the tourism cartoons selected for discussion have emerged from appraisal over several years, of an extensive range of bound volumes of comic magazines, illustrated comic tours, portfolios of single caricatures and comic postcard collections. The main thesis contained in this chapter is based on an analysis and discussion of an illustrative range of tourism cartoons, chosen on the basis that their content could be grouped into one of these four behavioural categories. The illustrations, therefore, which are included in this chapter, are typical examples of these behavioural categories.

This chapter follows an earlier paper on transport history in graphic satire (Seaton, 2011), and is part of a longer study that includes more psycho-social categories and a greater range of illustrations exemplifying each category.

Sensations and Perceptions

All visual images, no matter if these are drawn, painted, printed or sculpted, can represent the physical look of things, but they are not equally able to disseminate these appearances to large audiences. Graphic humour and other kinds of graphic reproduction could do so because they offered images in quantity. For every person who saw a drawing or painting, there were hundreds, later thousands and, later still, millions who were exposed to reproduced images. Thus it was that many more people became familiar with *the look* of travel and tourism conditions at home and abroad, including: the appearance of different horse drawn conveyances such as the French diligence, the Irish jaunting car, the English commercial travellers two wheel fly; the features of frontiers and terminals; the customs house, the city gate and the entrance to a harbour; the appearances of places, the traffic jams in metropolitan streets, the dilapidation of village inns and the grandeur of medieval facades.

But graphic humourists did more than sketch appearances. They could suggest the effects they had on mind and body, evoking the *sensations* and *perceptions* of touring. These had already become a feature of published tour memoirs and, to a lesser extent, of early guide books. But visual evidence was limited. Graphic satire brought travel sensations into pictorial focus on a scale that increased throughout the nineteenth century, suggesting what it felt like to travel, as well as the appearance of things seen. Texts with comic illustrations detailed the 'inconveniences' and 'miseries' of, for example, bad weather, poor roads, bilious sea crossings, coaches that were too slow or too fast, hard or flea-ridden beds, and the 'filth and nauseousness' of inns (Beresford, 1807, vol. 1, 96–128; vol. 2, 161–170). Figure 2.3 shows Woodward's ingenious, two-part depiction of the jolting and bumping of coach travel, and the ways in which it could be mitigated.

Figure 2.3 Symptoms of jolting and jolting preventives
Source: Woodward, 1796, p. 164.

This two-stage cartoon is only one of the many, brilliant comic inventions, made by Woodward in *Eccentric Excursions*, that offer evocative insights into the physicality of late eighteenth-century travel, that no painter ever came close to recording. The titles of other full page engravings in it suggest the range of tourist sensations he tackled: *The effects of rough pavements*, *Six different methods of holding a stick*, *Six of the most approved methods of appearing ridiculous on the ice*, *Stage coach passengers asleep*, *Riding with the wind* (which depicted an unfortunate traveller riding behind an overweight parson who had just farted), and, *Riding against the wind* (a sketch of a man riding his horse into a gale). Its combination of multiplicity and rubricity allowed graphic humour to deliver complex, physical cameos of the processes and progresses of touring.

It is a cliché, implicit in the concepts of sightseeing and the *tourist gaze*, that tourism privileged sight over other senses, putting visual perceptions at the centre of holiday behaviour (Urry, 1992, 2001; Urry and Larsen, 2011). Graphic humourists partly agreed and even extended the idea, by illustrating the number and variety of *new vantage points* created in the travelling process, the hotel rooms, ships cabins, railway carriages, clifftops, cable cars, castle towers which the tourist occupied, and the views they disclosed of people, places and things.

But cartoonists also interrogated the platitudes of stereotypical sightseeing and the *tourist gaze*, by revealing gaps between myths and reality; between what tourists expected, and what they actually experienced. The French artist Cham (1818–1879) (whose real name was Amadeus Noe) produced illustrated albums of comic tours that included sketches of England and Holland, which exposed the disappointments of some destination encounters. In one sketch, titled *Classic landscape on entering Holland*, he showed the reality to be a tangled, complex of railway junctions. Another sketch, with the deadpan title *View of the Thames at London*, showed the river barely visible under a gridlocked flotilla of identical boats (Cham, n.d., Plates 5 and 12). A more 'arty' exercise in the same vein was a cartoon featured in a comic tour of Scotland that poured gentle fun on Turner's celebrated painting *Rain, steam and speed*. This had been hailed as a miraculous experiment in impressionistically capturing the combined sublimity of nature and modern transport. The comic artist stole Turner's title, and attached it to a nondescript view of a flat-bottomed, little steam boat, called *The Linnet*, chugging out of Dunoon in the rain (R. Mc.E., 1869, p. 8).

In addition to disappointed expectations of place, graphic humourists exposed the self-delusions of tourists in the gap between claimed, and actual, tourist behaviour, something painting never attempted. A cartoon in *Judy* included a sketch of two lady ramblers, arriving belatedly at the summit of a peak, to find two young climbers who had forged on ahead, claiming they were eager to enjoy the sublime view, fast asleep (Judy, 1869). But perhaps the most instructive insight cartoonists provided into the validity of the *gaze* as a tourist attribute, was that it was frequently directed, not at scenery or indigenous populations, but at other tourists. This is an under-emphasised element in tourism theorizing. The human 'passing show' could be at least as absorbing as grand landscape, particularly in urban and resort destinations with high densities of visitors (Seaton, 2001). A hotel balcony was often depicted, less as a view on to nature and heritage, than a vantage point from which to observe the arrivals of interesting, new visitors. Cartoonists also suggested how the new places and spaces of tourism afforded opportunities for 'peeping Tom' moments when mischievous children, jealous lovers and other covert observants, spy, or listen in, on the developing, or developed, relations of others.

The Social Relations and Interactions of Tourism

Tourism was, and still is, a highly social activity. Most tourists travelled with people they knew and met others with whom they later formed acquaintance, friendship, or even came to love. They were also forced into contact with people they did not deliberately choose to know, including fellow travellers and guests, as well as a whole range of service employees who waited on them, directed them, fed them and managed their affairs. Comic artists delighted in exploring these social collisions, and where they took place. They represented the geography of the open air, such as the beaches, bays, river banks, mountainsides and hillsides, where specific social activities took place, each with its own rules of encounter, norms of fashion, situational accessories, and conventions of permissible proximity or distance. Comic albums abound in burlesqued representations of social life in the great outdoors: mishaps on ski slopes; arguments over bathing machines; picnics on mountain sides; parties sharing hair-raising experiences in open boats; and coping with unexpected, animal intrusions in the country.

Graphic humourists also explored the politics and mores of tourism's micro-spaces. These were unavoidable, infrastructural developments that threw people together, be it in railway's carriages, ship's dining rooms, hotels and boarding houses or museums and art galleries. Graphic humourists drew the welcome or unwelcome contacts such spaces produced, and ways in which they could be creatively manipulated by tourists to facilitate, or inhibit, further relations with those they did, or did not, wish to know better. Cameos of these infrastructural manoeuvres included: 'bun fights' in finding seats on coaches and trains; arguments over bathing machines; faux pas committed in art galleries or museums by the culturally challenged; cutting or cultivating acquaintance with other tourists while parading on pier or promenade; and maintaining family order and dignity in the congregation of English churches abroad.

But perhaps the most recurrent kind of social situation explored in graphic humour was holiday romance. New places meant new faces. This made tourism a sexually charged activity because it brought together eligible strangers. Spa towns and seaside resorts like Bath, Harrogate and Scarborough had always been potential dating agencies and marriage bureaus where partners might meet. The middle classes often travelled in large, family parties that might include new faces: an attractive, distant cousin; a school friend of one's brother; the daughter of a business partner of one's father. And even if the party did not include such eligible strangers, they might be found later among the other guests at popular resorts. Comic artists delighted in exploring the sexual possibilities of holidays including the conversational strategies used by young men to engage good-looking women, and women's ways of soliciting the attentions of men. They disclosed how seaside behaviour among the young could be a cattle market of mutual inspection, in which a telescope was for more than admiring a sea view, as this double cartoon discloses (see Figure 2.4).

The variety of settings and spaces in tourism not only made romantic encounters more likely but easier to pursue than in the chaperoned spaces of home. Phiz, the illustrator of Dickens, published an album showing how the young contrived

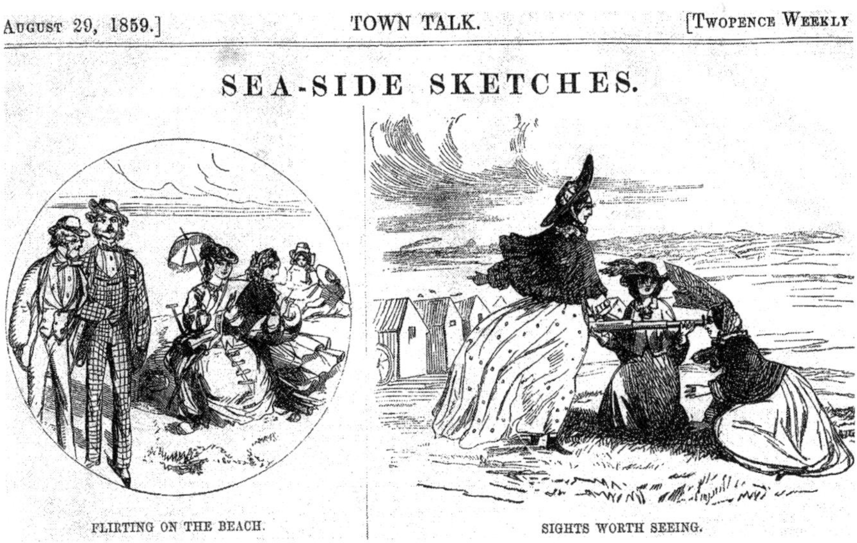

Figure 2.4 Flirting on the beach and sights worth seeing

Source: *Town Talk*, 1859.

to flirt and intrigue while the attention of accompanying relatives was turned to other things. Behind the back of elderly uncles giving tedious lectures on seaside geology or botany, couples whispered sweet nothings (Phiz, 1863). Half a century later postcards continually celebrated the possibilities of overt and covert flirting and holiday canoodling, at the seaside and in the country.

Graphic humour amused both men and women, although most of it was produced by men. A privileging of the male gaze was common and with it, recurrent stereotypes of female design and desire. A favorite trope was the spinster on holiday looking for a partner, or resentfully spying on younger, better-looking women, and seeking to spike their chances. On the rare occasions when women were published as graphic humourists they supported the stereotyping. A comic tour to the East, illustrated and written by Lady Dufferin, featured an aging spinster as heroine, whose social aspirations were only less than her determination to find a husband *en voyage* (Dufferin, 1863).

A perennial focus in tourism graphics was on the female body at the seaside. Mid-Victorian representations were of young, sylph-like women in beach wear that evoked baggy, Eastern trouser styles. *The Month* ran a summer series *Belles of the Month* which exhibited young beauties as individual subjects of the male gaze, in ways that anticipated beauty competitions, and *Playmates of the Month* in men's magazines a century later (The Month, 1851). Another magazine, *Town Talk* featured a cartoon, *Ocean Beauties*, which showed gatherings of beautiful young women playing, like mermaids in the sea in front of bathing machines (Town Talk, 1859, August 29). By the twentieth century comic postcard images featured raunchier representations of women, with less clothing, and voluptuous body shapes that elicited reactions of pop-eyed astonishment or delight, among young and old men around them.

Self-presentation, Identity and Performance

Seeing and being seen in the public, macro and micro spaces of tourism made holiday life continuous theatre, a potentially head-turning sequence of entrances and exits. Such performances required careful planning. Catholicism once included a concept known as *Ephodion* which meant spiritual preparation for a journey. Tourism came to have its own forms of secular *ephodion* in the industrial supply of commodified accessories available to the tourist. Advertisements in popular guide-books like Black's and Murray's, and women's fashion magazines, exemplify the growing, consumer culture of tourism. Part of it comprised functional items such as maps, branded luggage, travelling medical cabinets, portable desks and writing materials. But a large part was sartorial. Part of this was functional clothing serving the practical needs of protection from bad weather, the extreme heat or cold of foreign climates, or facilitating special activities such as climbing, hunting, shooting and fishing. These functional requirements resulted in specialist forms of footwear, protective head-gear, waterproofing, and thermal insulation, all of which were widely depicted in cartoons.

But there was also an extensive market in holiday wear that had no particular functional purpose. It was pure self-presentation, invented and promoted by the fashion industry for different holiday moments and occasions of the holiday: for walking out, for the theatre, for morning wear, for evening wear etc. Caricaturists tracked the narcissistic excesses of these *ephodionic* costume rituals, particularly, those for women. The image below (Figure 2.5), for example, was the main pull-out, plate to one of the annual Almanacs that *Punch* offered its readers at the end of the year (Leech, 1864, Plate 19).

One point of female self-presentation that attracted special attention from cartoonists in the 1850s was the vogue for long hair. At the seaside even middle class women began to let their hair down, for reasons that could variously be interpreted as the need to dry it, the pleasure of allowing the wind to run free through it, or the less admissible one of sensual exhibition. Graphic artists imagined some of the problems that could arise. One of these included the possibility that one's long hair could be exposed as false or even lost at sea (Fun, 1865, October 14), while another showed older women deploring the exhibitionism of young beauties with unbound tresses to die for (Fun, 1865, September 2).

Masculine self-presentation on holiday involved more than dress and external accessories. It was bound up with the more intangible requirements of being a 'good traveller'. The role was as keenly sought by the young male single tourist and the newly wedded husband on honeymoon with his wife, as by the paterfamilias travelling round Europe with a large family. Being a good traveller

Figure 2.5 Seaside fashions
Source: Leech, 1864.

required the maintenance of an equable, self-possession throughout the holiday, while assuming responsibility for women and children, ensuring their well-being, and paying their way. Recognition of the hazards and stresses to which tourism exposed its enthusiasts to, made the leadership implicit in 'the good traveller' role an important ideal for males to aspire to. But it was one that was difficult to maintain at all times. Cartoonists took delight in picking up cracks in the self-presentational ideal, and exposing the resulting shame and confusion of the flawed male, as in for example a merciless George Cruikshank cartoon (in Cruikshank, 1845) which depicts the humiliation of a man, wrongly dressed, and about to be laid low by sea sickness, who was unable to assume the role of 'good traveller' while those around him could. The narrative accompanying it also reveals that he is supposed to be the captain of the boat!

Since the tourist role involved public performance among people who didn't know each other, there were opportunities for identity concealment and deception, in the knowledge that the chances of having ones cover blown were low. Poses and impostures away from home were themes of Victorian fiction, as well as graphic humour. A common theme of caricature was the contrast between the Englishman at home and abroad. It was commonly constructed in side-by-side sketches that contrasted aristocratic pretension with the straightened circumstances that lay behind them. Another kind of stereotypical deception cartoonists exposed was that of the male fortune hunter on the make among silly women travellers. They sketched moments of truth when deceivers were unmasked as poor, rather than well-connected, married rather than single, and in one cartoon, the deceiver was discovered wooing other women by one who thought she had become his fiancée (Judy, 1869, October 6).

Tourism and Social Class

Self-presentation in tourism was linked to self-perceptions of status and social class. At home these were increasingly expressed by geographical segregation, with the rich and poor, living and working in different neighborhoods (Olsen, 1973). Tourism partly mirrored this geographical segregation. Different classes took their recreations in different places, thus reproducing a 'two nations' pattern of tourism behaviour, which writers such as Thomas Carlyle and Charles Dickens, observed was a growing feature of Victorian society.

But the differences were not simply geographical. Large swathes of the poor took no holidays at all, and when they did they were in England, not abroad and were Bank Holidays or half-holidays, that lasted no more than a day, not weeks or months; and their prevailing rules of engagement were alcoholic and carnivalesque. The day excursion and later the factory trip, were rare opportunities to 'let off steam' from a regimented life of drudgery in factory, workshop, retail or domestic service. Painting hardly touched places where the poor played, and most comic magazines, particularly *Punch*, mainly featured middle class, holiday behaviour. Only a few cartoonists stepped outside their bourgeois, comfort zones to visit the haunts of the excursion classes which included fun fairs, the races, Hampstead Heath, and trips by Thames boats or rail to Gravesend, Ramsgate and other day-out destinations. In June and July 1859 *Town Talk* ran a series, *Summer Days*, which comprised illustrated essays on packed, bus trips to Hampton Court, frolics on Hampstead Heath, and a visit to the little known pleasure ground of Windmill Hill, Gravesend. Six years later *Fun* ran an even more demotic series called, *The People's Pleasures*, which set out to investigate the 'little outings' of the classes that could not go abroad. It offered word-and-picture impressions of 'a steamboat up the river', Ramsgate Sands and a day excursion by rail and bus through the East End of Bethnal Green and Shoreditch. The series concluded with accounts of Barnet Fair and the fishing pubs of Temple Mills (Fun, 1865).

Social segregation did not always hold up on holiday. People from different social backgrounds sometimes found themselves in shared tourism spaces. Public transport, in particular, threw people together in ways that disrupted what Schivelbusch (1980) has called 'stimulus shields', the preserved social and physical barriers that normally regulated contact and separation between classes. Cartoonists made capital out of imagining the consequences of such occasions. A favorite situation that was regularly presented was that of cockneys and young 'oicks' exchanging impudent backchat with self-important travellers, attempting to maintain a dignified, social distance. A railway historian has noted how travel by train was particularly likely to offer such situations to graphic humourists:

> Caricaturists added to the stock of public fun by depicting grandees – or persons who thought themselves that – in trains, aloof and self-absorbed … (Simmons, 1991, p. 362)

Overall then, graphic humourists saw tourism as a social force that both reflected and occasionally subverted, class differences at home and abroad, but in general it was middle class mores that dominated comic magazines and tours. But there was one publication that bucked the general pattern.

For the last two decades of the nineteenth century a comic magazine costing a penny, called *Ally Sloper's Half Holiday* (see Figure 2.6), drew a large audience of lower class readers, with the antics of one whose carnivalesque vulgarity constantly prevailed over buttoned-up gentility. Its eponymous, red-nosed hero, Ally Sloper, was a boozy, social anarchist with a striking resemblance to the vintage film comedian, W. C. Fields, who dragged his long-suffering family around seaside resorts, society occasions and bank holiday excursions, creating mayhem among all comers, regardless of their social status. Every week his embarrassing exploits were featured in a full page cartoon, narrated in a language of genteel disapproval by his teenage daughter, Tootsie. This trail-blazing people's comic ran until the 1920s, and is today recognised, not just for the place it occupies in popular culture, but also for the graphic excellence of one of its artists, W. G. Baxter, who drew the covers during the 1880s but died aged only 32 in 1888.

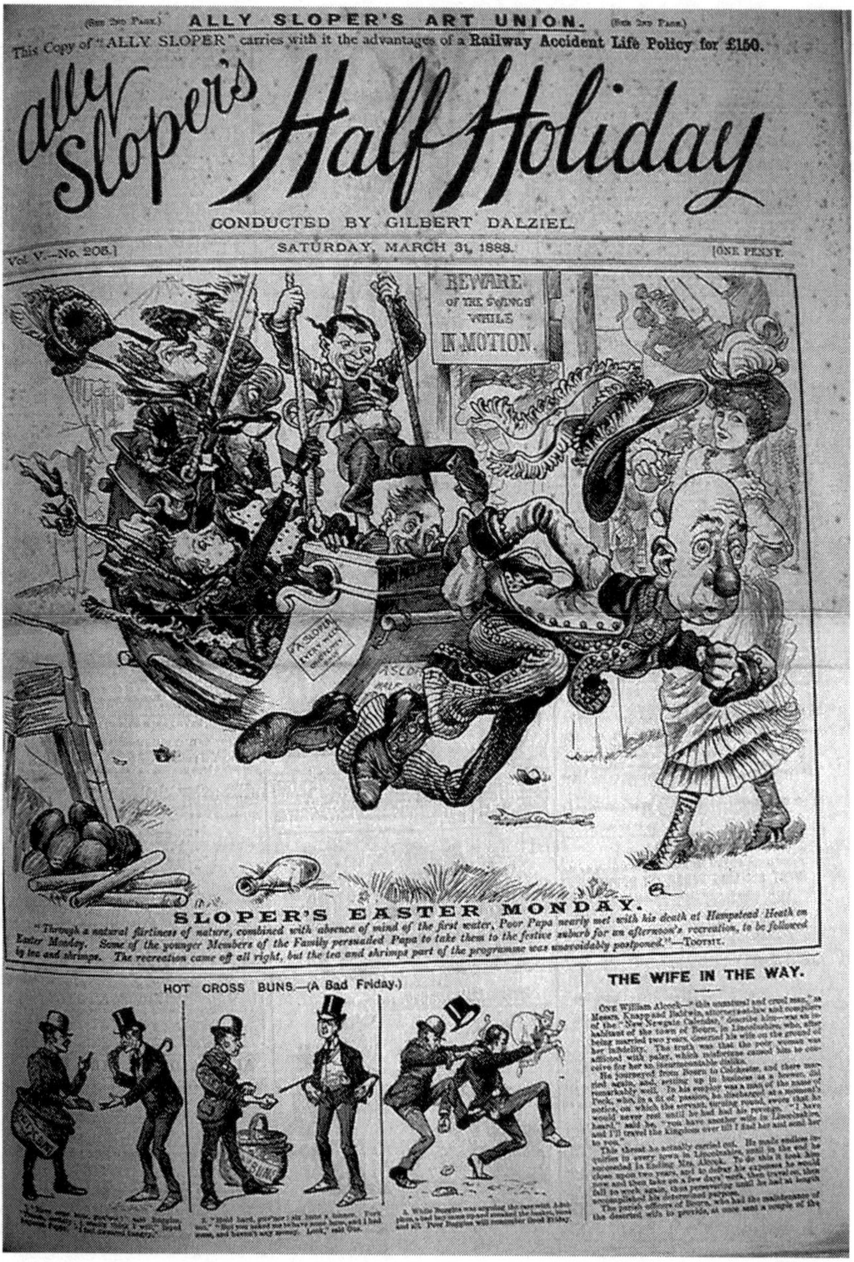
ALLY SLOPER'S ART UNION.
This Copy of "ALLY SLOPER" carries with it the advantages of a Railway Accident Life Policy for £150.
Ally Sloper's Half Holiday
CONDUCTED BY GILBERT DALZIEL.
Vol. V.—No. 205.] SATURDAY, MARCH 31, 1888. [ONE PENNY.

SLOPER'S EASTER MONDAY.
"Through a natural flirtiness of nature, combined with absence of mind of the first water, Poor Papa nearly met with his death at Hampstead Heath on Easter Monday. Some of the younger Members of the Family persuaded Papa to take them to the festive suburb for an afternoon's recreation, to be followed by tea and shrimps. The recreation came off all right, but the tea and shrimps part of the programme was unavoidably postponed."—Tootsie.
HOT CROSS BUNS.—(A Bad Friday.)
THE WIFE IN THE WAY.

Figure 2.6 ***Ally Sloper's Half Holiday***

Source: Easter Monday, 1888, March 1.

Conclusion

The aim of this chapter has been twofold. Firstly it has set out to establish the importance of graphic humour as *the pre-eminent* visual genre for understanding the psycho-social world of the tourist, during the period of tourism's expansion among the industrial, middle and working classes. Secondly, it has *exemplified* how the vast, resources of graphic humour in comic tours, comic magazines, caricatures and postcards can be *systematically* categorised under behavioural subject headings. Their individual cartoon contents were also qualitatively described and analysed, illuminating the phenomenal world of the tourist in history.

Though the graphic humour of travel has been featured incidentally in books on caricature as a form (for example, George, 1967), and also appeared as ad hoc illustration in historical accounts of tourism (Löschburg, 1979), there has been no recognition of its unique, quantitative and qualitative superiority to all other visual art forms, as a systematic lens through which to apprehend the behavioural world of the tourist. Painting was ideologically sealed off from representing tourism by a patronage system in which neither artists, nor their sponsors, deemed its behavioural effects worth recording. More importantly, painting was disabled from representing the temporal and spatial mobilities of travel by its inherent limitations as a medium, laborious to use, and static in its effects. It was not just aesthetic choice, but practical necessity, that led to the canonical ideology that painting's quintessential effects lay in representing the solemn, immobilities of still life, comprising not just fruits and flowers, but posed portraits, religious tableaux, and landscape views. 'Art', as Gombrich observed, '… [had] to compensate for *the loss of the time dimension* by *concentrating* all required information into one *arrested image*' (1960, p. 345, italics added).[2]

By contrast, graphic humour could, and did, represent the physical and psycho-social mobilities of tourism in multiple, abbreviatory sketches, conveying transformations of place and time, with meanings that could be punched home by the allowable rubricity of the cartoon form.

References

Alderson, F. (1970). *The Comic Postcard in English Life*. London: David and Charles.

Ally Sloper's Half Holiday 1884–1916 [a journal].

Atherton, H. M. (1974). *Political Prints in the Age of Hogarth*. Oxford: Clarendon.

2 Gombrich was writing about caricature and comparing it with the more overtly 'serious' business of painting, in relation to which he noted that the dignified and static effects of painting suited the self-presentational preferences of sponsors who did not 'do' laughter or emotion. 'The noble' he asserted, 'neither laugh nor cry' (Gombrich, 1960, p. 349). Hence, the marginalisation of caricature and graphic humour that this chapter seeks to reverse as a historical behavioural resource.

Atkinson, J. P. (1878). *A Week at the Lakes. The Adventures of Mr Dobbs*. London: Macmillan.
Bailey, P. (1978). *Leisure and Class in Victorian England: Rational Recreation and the Contest for Control 1830–1885*. London: Routledge and Kegan Paul.
Bakhtin, M. M. (1984). *Rabelais and His World*. Bloomington: Indiana University Press.
Barker, R. (1988). *Naughty Seaside Postcards*. London: Penguin.
Beresford, J. (1806–1807). *The Miseries of Human Life or the Groans of Samuel Sensitive and Timothy Testy*. London: Miller.
Buckland, E. (1984). *The World of Donald McGill*. London: Blandford.
Calder-Marshal, A. (1966). *Wish you Were Here*. London: Hutchinson.
Cham. (n. d. 1850s). *Impressions Lithographiques de Voyage*. Paris: Chez Aubert.
Cruikshank, G. (1845). *George Cruikshank's Table Book* (edited by Beckett, G. À.). London: Punch Office.
Cunningham, H. (1980). *Leisure in the Industrial Revolution 1780–1880*. London: Croom Helm.
Doyle, R. (1854). *The Foreign Tour of Messrs Brown, Jones and Robinson*. London: Bradbury and Evans.
Dufferin, L. (1863). *Lispings from Low Latitudes; Or Extracts from the Journal of the Hon. Impulsia Gushington*. London: John Murray.
Eldon, A. (1828). *The Continental Travellers Oracle or Maxims for Locomotion*. London: Henry Colburn.
Everitt, G. (1888). *English Caricaturists and Graphic Humourists of the Nineteenth Century. How They Illustrated and Interpreted Their Times*. London: Swan Sonnenschein, Le Bas and Lowrey.
Fosbroke, T. D. (1826). *The Tourist's Grammar or Rules relating to the Scenery and Antiquities Incident to Travellers*. London: John Nichols and Son.
Fun 1861–1901 [a journal].
George, D. (1959). *English Political Caricature 1793–1832*. Oxford: Clarendon Press.
George, D. (1967). *Hogarth to Cruikshank: Social Change in Graphic Satire*. New York: Walker and Company.
Gombrich, E. H. (1960). *Art and Illusion*. London: Phaidon Press.
Grand-Carteret, J. (1927–1928). *L'Histoire, la Vie, les Moeurs et la Curiosité par l'Image, le Pamphlet et le Document 1450–1900*. Paris: Librarie de la Curiosité et des Beaux-Arts.
Green, B. (1976). *I've Lost my Little Willie*. London: Elm Tree/Arrow Books.
Grego, J. (1880). *Rowlandson the Caricaturist. A Selection from his Works with Anecdotal Descriptions of his Famous Caricatures*. London: Chatto and Windus.
Hogarth, W. (1952). *Hogarth's Peregrination* (ed. by C. Mitchell). Oxford: Oxford University Press.
James, L. (1976). *Print and the People*. London: Allen Lane.
Judy 1867–1907 [a journal].

Kitchiner, W. (1827). *The Traveller's Oracle, or, Maxims for Locomotion*. London: Henry Colburn.

Klingender, F. D. (1943). *Hogarth and English caricature*. London: Transatlantic Arts.

Kunzle, D. (1973–1994). *The Early Comic Strip*, 2 vols. Los Angeles: University of California Press.

Leech, J. (1864). *Follies of the Year*. London: Bradbury and Evans.

Löschburg, W. (1979). *A History of Travel*. Leipzig: Interdruck, Graphischer GrosBbetrieb.

Mayhew, M. and Cruikshank, G. (1851). *1851: Or the Adventures of Mr. and Mrs. Sandboys and Family who Came up to London to see the Great Exhibition*. London: David Bogue.

Olsen, D. (1973). Victorian London: Specialisation, Segregation and Privacy. *Victorian Studies*, March, 264–278.

Orwell, G. (1941). The Art of Donald McGill. *Horizon*, September.

Paulson, R. (1865). *Hogarth's Graphic Works*. New Haven and London: Yale University Press.

Phiz (Hablot K. Browne) (1863). *Sketches of the Seaside and the Country*. London: Graphotyping Company.

R. Mc. E. (1869). *My Hearts in the Highlands. Extravagances Suggested by a Sketching Holiday in the North*. Glasgow: David Bryce.

Rowlandson, T. (1812). *The Tour of Dr. Syntax in Search of the Picturesque*. London: Ackermann.

Schivelbusch, W. (1980). *The Railway Journey*. Oxford: Blackwell.

Seaton, A. V. (2001). *Traveller Toffs and Tourist Scum*. Paper presented at the Locating the Victorians Conference, The Science Museum, London.

Seaton A. V. (2011). The psycho-social impacts of Travel and Transport developments in graphic satire 1780-1850: an exhibition agenda. Paper presented at the Transport and Mobility on Display, 9th International Conference of the International Association for the History of Transport Traffic and Mobility, October 6-9, Deutsches Technikmuseum, Berlin.

Simmel, G. (1950). *The Sociology of Georg Simmel* (K. H. Wolff, Trans.). Glencoe: Free Press.

Simmons, J. (1992). *The Victorian Railway*. London: Thames and Hudson.

Staff, F. (1966). *The Picture Postcard and its Origins*. London: Lutterworth Press.

The Month 1851 [a journal].

Town Talk 1855–1857 [a journal].

Towner, J. (1996). *A Historical Geography of Recreation and Tourism in the Western World 1540–1940*. London: John Wiley.

Uglow, J. (1997). *Hogarth: A Life and a World*. London: Faber and Faber.

Urry, J. (1990). *The Tourist Gaze*. London: Sage.

Urry, J. (2002). *The Tourist Gaze* (2nd ed.). London: Sage.

Urry, J. and Larsen, J. (2011). *The Tourist Gaze 3.0*. London: Sage.

Woodward, G. M. (1796). *Eccentric Excursions, or, Literary and Pictorial Sketches of Countenance, Character and Country in Different Parts of England and South Wales*. London: Allen and West.
Wright, T. and Evans, R. H. (1851). *Historical and Descriptive Account of the Caricatures of James Gillray*. London: Bohn.
Wykes, A. (1977). *Saucy Seaside Postcards*. London: Jupiter.

Chapter 3
'Lofty Hills Streaming With Water-Falls': The Tourist Landscape of Johnson and Boswell

Brian Hudson

Introduction

Waterfalls have long fascinated travellers and artists. Scholars, too, have given much attention to these popular landscape features. Today, many articles on waterfalls can be found in journals of geography, earth science, landscape and tourism studies. Books on waterfalls have been published since the nineteenth century. Two notable examples are *Niagara and Other Famous Cataracts of the World* (Holley, 1883), and *Great Waterfalls, Cataracts and Geysers* (Gibson, 1887). Rashleigh's (1935) *Among the Waterfalls of the World* remains a classic in the field. *Waterfall: Nature and Culture* (Hudson, 2012) is a recent addition to the literature.

Since the 1980s, waterfall guidebooks have proliferated around the world. Among the many national and regional waterfall books that have been published, Louis Stott's (1987) *The Waterfalls of Scotland Worth Gaun a Mile to See* is possibly unequalled in terms of systematic detailed observation, comprehensiveness and breadth of approach. It deals with a wide variety of topics ranging from locational, geological and statistical information to historical and literary associations. In his book, Stott mentions Samuel Johnson and James Boswell, who, together, toured Scotland in 1773, observing many waterfalls on their journey. The following examination of the Scottish tour undertaken by Johnson and Boswell is offered as a contribution to the study of the experience of waterfalls and their representation in the world of art.

Waterfalls in the Art World

This chapter examines works by two eminent eighteenth-century British writers in order to throw light on the aesthetic response to waterfalls, landscape features which have, for millennia, impressed travellers and inspired artists. The artistic representation of waterfalls has a long history. As Rashleigh expressed it, 'a thousand painters and poets all down the centuries have caught their message and have been inspired by it' (1935, p. 14). Other art forms, too, ranging from music to landscape architecture, have taken inspiration from waterfalls which have evoked feelings of delight and awe since ancient times (Hudson, 2012). Dating from the third century BC, the first of the *Idylls* by Greek poet Theocritus begins with a scene beside a waterfall, the sound of which is compared to the music played by a shepherd on his pipe.

In European literature and painting, waterfalls began to feature commonly from the seventeenth century, becoming very popular subjects during the late eighteenth and early nineteenth centuries. This was the period of the Grand Tour when wealthy young travellers went abroad to gain knowledge and experience of manners and cultures in foreign countries. At first they visited mainly places of cultural significance, such as great cities, famous buildings, classical ruins and important art collections, but later, as the taste for landscape beauty became fashionable, more tourists began to travel to wild mountainous regions, places where rivers and streams often form spectacular waterfalls. These increasingly became subjects of European painting and poetry, works that further encouraged travel to see places thus celebrated. Among the artists who took inspiration from waterfalls were painters Francis Towne (1739–1816), Philip de Loutherbourg (1740–1812), Thomas Girtin (1775–1802), and Joseph Turner (1775–1851). Goethe (1749–1832), Wordsworth (1770–1850) and Southey (1774–1843) are perhaps the best known poets whose works include famous pieces about waterfalls. Waterfalls have also played a significant role in the art of many different cultures around the globe. The remarkable popularity of waterfalls in Chinese culture is exemplified in its poetry and painting dating back a thousand years and more.

Such is the appeal of tumbling rivers and streams that for centuries people with the inclination, time and means to enjoy the pleasures of landscape have travelled far, often experiencing discomfort, even danger, in order to visit waterfalls (Hudson, 2001). It is not surprising, therefore, that waterfalls feature prominently in travel literature. In a book on travel, tourism and art, it is important to emphasize here that, ‘Unlike our mediocre guide books, Georgian voyages and travels were expected to be serious intellectual documents with important information to contribute to the advancement of the arts and sciences’ (Curley, 1976, p. 47–48). The books themselves were art works that sometimes achieved high standards in terms of their literary merit and their illustrations. In his discussion of Samuel Johnson’s *A Journey to the Western Islands of Scotland*, Curley describes this ‘elegant travel book’ as a ‘skillful blend of philosophic art and travel’ (1976, p. 183). Johnson’s companion on his Scottish tour was James Boswell who also wrote a book about this journey, having previously published accounts of his travels in Europe. Of Johnson’s fellow traveller, Carey McIntosh wrote, ‘It should be clear that if appropriateness and variation may count as criteria for the artistry of a writer’s style, Boswell gets very high marks on both counts’ (1995, p. 152). Johnson and Boswell are recognized masters of the genre of travel writing, and it is the response of these great men of letters to the Scottish landscape, particularly the waterfalls they saw there, that is the subject of the following discussion.

The Motivation for the Scottish Tour

Englishman Dr. Samuel Johnson and his Scottish friend and biographer James Boswell toured Scotland together from August until November 1773. At that time, Johnson was 63 years old, Boswell about half his companion's age. Johnson wrote, 'I had desired to visit the *Hebrides*, or Western Islands of Scotland, so long, that I scarcely remember how the wish was originally excited' (1944, p. 3). It seems very likely, however, that a book given to youthful Samuel by his father, *A Description of the Western Islands of Scotland* by Martin Martin (1703), sowed the seed of curiosity which eventually led to Johnson's decision to see Scotland for himself (Boswell, 1964, p. 13). This book, written by a native of Skye who travelled among the Western Islands of Scotland in the late seventeenth century, also stimulated Boswell's enthusiasm for the journey:

> Martin's account of those islands had impressed us with the notion that we might contemplate a system of life almost totally different from what we had been accustomed to see; and, to find simplicity and wildness, and all circumstances of remote time or place, so near to our native great island, was an object within reach of reasonable curiosity. (Boswell, 1964, p. 13)

References to their Scottish tour in Johnson's and Boswell's accounts indicate that the men were familiar with the works of other, later authors who had published descriptions of their journeys in Britain, notably Welsh naturalist and antiquary Thomas Pennant (1726–1798) and English writer on agriculture, economics and social statistics Arthur Young (1741–1820). In their writings, both Pennant and Young expressed a fascination with waterfalls that was typical of their time. The plates in Pennant's (1771) *A Tour in Scotland 1769* include a couple illustrating waterfalls, one of them the 'Upper Fall of Fyers' spanned by a precarious looking bridge made of tree trunks and sods (See Figure 3.1). The Falls of Foyers, as they are known today, were among the famous sights of the Scottish Highlands that Johnson and Boswell made a particular effort to see. Indeed, Boswell notes that waterfalls were among the 'wild objects' that Johnson especially wanted to see on his Scottish tour (Boswell, 1964, p. 112).

Johnson's attitude to travel is best understood in the light of his belief that a knowledge of geography is necessary for a proper understanding of the human condition, and that, while written accounts are important sources of information, first hand experience is more reliable (Curley, 1976). Inevitably, we must rely on the descriptions of others for much of our knowledge of the world, but we should remember that these are always incomplete and possibly unreliable. Moreover, whatever one may feel about the wilder parts of the Earth, they deserve our attention because they occupy such a great part of the world in which we live. Johnson observed that:

Figure 3.1 Eighteenth-century print of the upper Falls of Foyers

Source: Originally published in Pennant, 1771; digital image source: <http://www.bun.kyoto-u.ac.jp/~hashida/MAEWWW/engra/pictures/fyers2.jpg>.

> Regions mountainous and wild, thinly inhabited, and little cultivated, make a great part of the earth, and he that has never seen them, must live unacquainted with much of the face of nature, and with one of the great scenes of human existence. (Johnson, 1944, p. 35)

It was Johnson's curiosity about the wilder parts of Scotland that enticed him to make the long journey north, despite his dislike for protracted separation from 'civilized life' (Boswell, 1964, p. 208). Johnson's serious purpose as a traveller should not disguise the fact that he was a tourist in the sense that he had the desire to enjoy pleasurable new experiences away from his everyday environment, and who, having the necessary leisure and means, used them to do so. Boswell clearly recognizes this when describing Johnson riding through the Highlands 'on horseback, jaunting about at his ease in quest of pleasure and novelty, the very different occupations from his former laborious life' (Boswell, 1964, p. 132). Johnson the literary artist is revealed in his written descriptions which vividly bring to life the scenes and events that he experienced on his travels. Johnson, far more than Boswell, provides the reader with a word picture of the Scottish landscape, which he often describes in terms of the Sublime.

The Sublime, the Beautiful and the Picturesque

The aesthetic concepts of the sublime, the beautiful and the picturesque were subjects of much intellectual discussion at that time. In the opinion of Andrews, Johnson 'had little of that taste for the sublime which Scottish scenery above all could satisfy' (1989, p. 200), but Johnson's account of his journey to the Hebrides, and his later one, which records a tour to parts of England and Wales, provide ample evidence that this eminent man of letters shared the then fashionable taste for sublime, beautiful, and picturesque landscapes. Geographer Robert Mayhew is a more recent scholar who draws attention to Johnson's use of the language of the beautiful, sublime and picturesque (Mayhew, 2004, p. 244). Notable among those who contributed to the discussion of these aesthetic ideas was a friend of Johnson, Edmund Burke. Burke's influential work *A Philosophical Enquiry into the Origin of Our Ideas of the Sublime and Beautiful* was first published in 1756, the second, greatly enlarged edition appearing in 1757. It was around this time, too, that William Gilpin was developing and publishing his ideas on the picturesque which he introduced in his popular *An Essay on Prints* (Gilpin, 1768).

According to Burke, the characteristics of the beautiful include gradual variation, smoothness, smallness, lightness and delicacy, while the sublime is associated with abruptness, vastness, ruggedness, massiveness, darkness and gloom. The former is related to pleasure, the latter to pain (Burke, 1757). We experience the sublime when we feel a 'delightful horror' at the prospect of danger although, in fact, we are detached or protected from any real danger (Sharma, 1996, p. 447–462). A picturesque scene is, quite literally, one that meets the eye as a satisfying picture, a composition which may combine elements of both the beautiful and the sublime. Characteristics of picturesque taste include a preference for rough textures, irregularity and partial concealment, and among the objects commonly found in picturesque scenes are ancient ruins, rustic bridges, quaint cottages, mills and grottoes. Occasionally, Johnson and Boswell use the word 'romantick' when describing a scene. Among the meanings that Johnson gives to this word in his famous dictionary are 'wild' and 'wild scenery'. In the literature of the Romantic age, waterfalls were commonly described in terms of the beautiful, the sublime and the picturesque, and, depending on circumstances, a waterfall may possess characteristics that combine more than one of these qualities (Hudson, 2000).

The Landscape Tastes of Johnson and Boswell

Neither Johnson nor Boswell showed a passionate interest in landscape beauty, and bleak moors and mountains held little appeal for them. Commenting on Johnson's attitude to landscape, Boswell wrote, 'I have a notion he at no time had much taste for rural beauties. I have myself very little' (Boswell, 1964, p. 112). At least in his younger days, Boswell was able to appreciate and describe some rural landscapes and although his writings are mainly concerned with people, he records the occasional observation on the landscape he sees on his travels. Commenting on Boswell's travels in Europe, Perreton observes, 'during his continental tour, one sees a Boswell whose reaction to landscape is as sophisticated as that of most young gentlemen on the tour' (1995, p. 37).While travelling in Switzerland, for example, he records 'We had on all hands grand mountains and rocks and now and then a castle' (Boswell, 1953, p. 204). Later, while he was waiting to pay a visit to Rousseau in the Swiss Jura village of Môtiers, Boswell

> ... strolled pensive by the side of the river Reuse in a beautiful wild valley surrounded by immense mountains, some covered with frowning rocks, others with clustering pines, and others with glittering snow. The fresh, healthful air and the romantic prospect around gave me a vigorous and solemn tone. (Boswell, 1953, p. 215)

Even on his Scottish tour, Boswell reveals a taste for romantic scenery when he describes a spot where a ruined castle stands on a rocky, wooded site overlooking a river. He remarks, 'I cannot figure a more romantick scene' (Boswell, 1964, p. 379).

Despite Boswell's comment on Johnson's and his own lack of great enthusiasm for rural landscapes, it is evident that both men could derive pleasure from scenic beauty. In Curley's opinion, Johnson 'could respond to sublime scenery as strongly as any nature enthusiast but preferred an admixture of elegance and sublimity on moral and psychological grounds. Sublime landscapes implied sturdy virtues but also suggested poverty, chaos, and human helplessness' (Curley, 1976, p. 213). While understanding the difference between the beautiful and the sublime as expounded by Edmund Burke, Johnson, nevertheless, recognized 'the ideal merging of the two' (Curley, 1976, p. 102). Johnson reveals a taste for the beautiful in a letter he wrote while touring Wales in 1774: 'Wales, so far as I have yet seen it, is a very beautiful and rich country, all enclosed and planted' (Johnson, 1816, p. 92). His preference for tamed, cultivated landscapes is clearly shown in the following quotation from his Scottish journal:

> An eye accustomed to flowery pastures and waving harvests is astonished and repelled by this wide extent of hopeless sterility. The appearance is that of matter incapable of form or usefulness, dismissed by nature from her care, and disinherited of her favours, left in its original elemental state, or quickened only with one sullen power of useless vegetation. (Johnson, 1944, p. 34–35)

The Search for the Picturesque (Andrews, 1989) that attracted many visitors to Scotland in the eighteenth and nineteenth centuries, was clearly not the principal reason for the decision by Johnson and Boswell to tour that country. Nevertheless, mountains and waterfalls, features that appeal to lovers of picturesque landscapes, were among the things that they wanted to see. As Boswell explained,

> Dr. Johnson ... always said, that he was not come to Scotland to see fine places, of which there were enough in England; but wild objects, – mountains, – waterfalls, – peculiar manners; in short, things he had not seen before. (Boswell, 1964, p. 112)

Waterfalls were among the curiosities and beauties of nature that were particularly popular with tourists in the Romantic era. Johnson's comments on Highland scenery reveal that he appreciated cascading streams as welcome sights in what was, for him, a generally dull landscape. As quoted above, Boswell specifically mentions waterfalls as things he believed his travel companion wanted to see on their Scottish tour (Boswell, 1964, p. 112). Johnson clearly found pleasure in waterfalls which are frequently mentioned in his accounts of his Scottish and later Welsh tours. The multifarious aesthetic appeal of waterfalls has been analysed elsewhere (Hudson, 2000, 2012). In this chapter, the responses of Johnson and Boswell to these landscape features are considered mainly in terms of the beautiful, the sublime and the picturesque, concepts widely discussed at the time of their tour.

Waterfalls Observed by Johnson and Boswell

Among the most famous British waterfalls in the late eighteenth century were England's High Force on the River Tees, Pistyll Rhaeadr in Wales, and the Falls of Clyde and Falls of Foyers in Scotland. On their Scottish tour, Johnson and Boswell did not visit the grand series of falls on the Clyde but, escorted by guides, they did make an excursion to 'the celebrated fall of Fiers' near Loch Ness (Johnson, 1944, p. 29). It is possible that Pennant's description of this 'vast cataract' where 'the foam, like a great cloud of smoke, rises and fills the air' (1771, p. 170) helped to make this sight one not to be missed by the two tourists. Johnson's account of this side trip is printed under a bold heading, 'Fall of Fiers', followed by a lengthy description part of which is reproduced below:

> We desired our guides to shew us the fall, and dismounting, clambered over very rugged crags, till I began to wish that our curiosity might have been gratified with less trouble and danger. We came at last to a place where we could overlook the river, and saw a channel torn, as it seems, through black piles of stone, by which the stream is obstructed and broken, till it comes to a very steep descent, of such dreadful depth, that we were naturally inclined to turn aside our eyes.
>
> But we visited the place at an unseasonable time, and found it divested of its dignity and terror. Nature never gives everything at once. A long continuance of dry weather, which made the rest of the way easy and delightful, deprived us of the pleasure expected from the fall of Fiers. The river having now no water but what the springs supply, showed us only a swift current, clear and shallow, fretting over the asperities of the rocky bottom, and we were left to exercise our thoughts, by endeavouring to conceive the effect of a thousand streams poured from the mountains into one channel, struggling for expansion in a narrow passage, exasperated by rocks rising in their way, and at last discharging all their violence of waters by a sudden fall through the horrid chasm. (Johnson, 1944, p. 29)

In this long description of a famous waterfall, Johnson demonstrates his skill as a portrayer of landscape through the medium of words, the picture he creates comparable, in its use of sublime imagery, with the paintings of waterfalls by contemporary artists such as de Loutherbourg and Towne. We recognize Johnson's familiarity with the ideas and vocabulary of the sublime in his use of terms and phrases such as 'rugged crags'; 'black piles of stone'; 'of such dreadful depth, that we were naturally inclined to turn aside our eyes'; 'dignity and terror'; 'discharging all their violence of waters'; 'sudden fall'; and 'horrid chasm'. Ruggedness, blackness, abruptness, terror, horror, dread, violence and threatening size are all qualities that are typically associated with sublimity. Later in his journal, Johnson comments on 'two pleasing water-falls' that he saw near Dunvegan Castle on the Isle of Skye where he also observed 'lofty hills streaming with water-falls' (Johnson, 1944, p. 62 and 68). In light of Burke's remarks on aesthetics, pleasure and pain, the word 'pleasing' is more suggestive of the concept of the beautiful than the sublime.

With the onset of stormy weather back on the Scottish mainland, the spectacle of numerous waterfalls provided some aesthetic compensation for the discomforts of wind and rain. As Johnson noted, 'In this rainy season the hills streamed with waterfalls' (1944, p. 144). It was not only the sight of the tumbling streams that impressed the traveller. In the darkness following nightfall, 'we could discern the cataracts which poured down the hills … The wind was loud, the rain was heavy, and the whistling of the blast, the fall of the shower, the rush of the cataracts and the roar of the torrent, made a nobler chorus of rough musick of nature than it had ever been my chance to see hear before' (ibid, 143–144). Clearly, Johnson was sensitive to the aesthetic appeal of both the sight and sound of waterfalls and had the usual preference for seeing them in spate rather than when reduced in volume by drought (Hudson, 2002). This latter view is corroborated by an observation in the Dunvegan section of Boswell's *Journal of a Tour to The Hebrides*: 'It was still a storm of wind and rain. Dr. Johnson however walked out with M'Leod, and saw Rorie More's cascade in full perfection' (Boswell, 1964, p. 215).

Later in his *Journal*, Boswell refers to other waterfalls on Skye. In one place '… there runs a rivulet with a good quantity of water, forming several cascades, which make a considerable appearance and sound' (Boswell, 1964, p. 218). Elsewhere, Boswell makes another reference to the sound made by a waterfall, one within earshot of Dunvegan Castle where the two travellers were guests of the McLeods.

> This room had formerly been the bed-chamber of Sir Roderick Macleod, one of the old Lairds; and he chose it, because, behind it, there was a considerable cascade, the sound of which disposed him to sleep. (Boswell, 1964, p. 207–208)

Boswell's use of bland phrases such as 'a rivulet with a good quantity of water', and 'a considerable cascade' contrast strongly with the descriptive power of Johnson's language. This reflects Boswell's 'concern for people in the landscape rather than the natural world as such' (Mayhew, 2004, p. 240). While staying at

Talisker House on Skye, Boswell went on a short excursion, this time without Johnson, which the biographer remembered particularly for the waterfalls seen along the way: 'Talisker, Mr. M'Queen, and I, walked out, and looked at no less than fifteen different water-falls near the house, in the space of about a quarter of a mile' (Boswell, 1964, p. 254). Boswell records one more waterfall seen on his tour with Johnson, this time at Mackinnon's Cave on the Isle of Mull:

> Upon the left of its entrance there is a cascade, almost perpendicular from the top to the bottom of the rock. There is a tradition that it was conducted thither artificially, to supply the inhabitants of the cave with water. Dr. Johnson gave no credit to this tradition. (Boswell, 1964, p. 331)

Most remarkable about Boswell's record of his visits to waterfalls while on tour in Scotland with Johnson is his omission of any mention of the Fall of Foyers which his companion described fully in his account of the journey. It could be argued that Johnson's detailed description made another account of the event unnecessary, but Boswell did not omit another event on the same day which Johnson also described at length – a visit to an old woman's primitive cottage beside Loch Ness. Here we may have an example of Boswell's greater interest in people than in the physical landscape they inhabit. Perhaps, too, as a patriotic Scotsman, he was reluctant to record the disappointing sight of the famous Scottish waterfall in its least impressive state, with the stream much reduced by drought. Elsewhere in his *Journal*, however, he records his disappointment with places he visited:

> I must own too that the old castle and cave, like many other things, of which one hears much, did not answer my expectations. People are every where apt to magnify the curiosities of their country. (Boswell, 1964, p. 173)

We can discern, in the sentence preceding the quotation above, another possible reason for the surprising omission. He writes, 'I find a great difficulty in describing visible objects' (Boswell, 1964, p. 173). Later, when he records a visit to an old ruin in a Highland setting, he admits:

> … I had endeavoured, in my journal, to state such particulars as might give some idea of it, and of the surrounding scenery; but from the great difficulty of describing visible objects, I found my account so unsatisfactory, that … I have omitted it. (Boswell, 1964, p. 219)

The reader may notice that, although Boswell makes several references to waterfalls in his *Journal*, he provides little description of them. Unlike Boswell, Johnson demonstrates powerful descriptive skills when he records his observations of landscape, including waterfalls. One of Johnson's most dramatic landscape descriptions, referring to the setting of Dunvegan Castle on the Isle of Skye, is recorded by Boswell:

> It looks as if it had been let down from heaven by the four corners, to be the residence of a Chief … it has many circumstances of natural grandeur, suited to the seat of a Highland chief; it has the sea, – islands, – rocks, – hills, – a noble cascade … (Boswell, 1964, p. 223)

The sentiments expressed here may have been coloured by Johnson's wish to be gracious to a kind host, but it is evident that the speaker is sensitive to features that combine to make a magnificent natural landscape, not least noble, or sublime, waterfalls.

Conclusion

Johnson and Boswell toured Scotland at a time when travel for the aesthetic enjoyment of landscape was becoming increasingly fashionable. Among the landscape features that appealed most strongly to travellers in search of the beautiful, the sublime and the picturesque were waterfalls. Many of these became famous through the work of travel writers, poets and painters who shared and popularized the aesthetic tastes of the period. By examining the writings of Johnson and Boswell, particularly the descriptions of waterfalls seen on their Scottish tour, this chapter contributes to the understanding of the role the concepts of the beautiful, sublime and picturesque played in eighteenth-century British literary arts.

This chapter has shown that, despite Boswell's assertion that neither he nor Johnson 'had much taste for rural beauties', and their lack of enthusiasm for the bleaker terrain over which they travelled, both men reveal in their writings that they derived considerable pleasure from landscape beauty. They had some appreciation of 'romantic' scenery. Although more interested in people than places, Boswell occasionally expressed delight in the landscapes he saw on the Continent and in Scotland, and Johnson is clearly enthusiastic in some of his descriptions of scenery he observed on his British journeys. Johnson, in particular, shows considerable powers of observation and ability to use language as a medium for creating images of landscape. He reveals a taste for the sublime and the beautiful, and his descriptions of scenery show that he was aware of the difference between these two aesthetic concepts.

Both men took particular pleasure in the sight and sounds of waterfalls, landscape features which were among the sights the two travellers set out to see on their tour of Scotland. Even in the wildest country during periods of stormy weather, the spectacle and roar of cascading streams provided the travellers with aesthetic compensation for their physical discomfort. One of the great disappointments of the tour was the men's visit to the Falls of Foyers, a waterfall already famous by that time, but which failed to meet expectations because prolonged drought had much reduced its flow. Nevertheless, Johnson devoted considerable space to its description in his journal, his choice of words and imagery revealing his familiarity with the concept of the sublime.

Boswell's omission to mention the Falls of Foyers in his own account of the Scottish tour may reflect his stated lack of confidence in being able to describe visible objects, including scenery, to his satisfaction. There are several references to waterfalls in Boswell's journal, but very little description of them. In contrast, Johnson devotes much of his journal to the description and appraisal of the landscape, giving considerable attention to waterfalls, natural features that clearly gave him particular delight. In the art of travel writing, Johnson was a master of landscape portrayal, his friend and companion Boswell being at his literary best with accounts of people and their conversation. Although neither writer admitted to a great love of landscape, both reflected the aesthetic tastes of their era by showing an appreciation of scenic beauty, including a fascination with waterfalls.

References

Andrews, M. (1989). *The Search for the Picturesque: Landscape Aesthetics and Tourism in Britain, 1760–1800.* Aldershot: Scolar Press.

Boswell, J. (1953). *Boswell on The Grand Tour: Germany and Switzerland 1764* (edited by Frederick A. Pottle). London: McGraw Hill Book Co.

Boswell, J. (1964). *Boswell's Life of Johnson Together with Boswell's Journal of a Tour to The Hebrides and Johnson's Diary of a Journey to North Wales* (edited by George Birkbeck Hill), *Volume V: The Tour of the Hebrides and The Journey Into North Wales* (2nd ed.). Oxford: Clarendon Press.

Burke, E. (1757). *A Philosophical Enquiry Into the Origin of Our Ideas of the Sublime and Beautiful.* London: Printed for R. and J. Dodsley.

Curley, T. M. (1976). *Samuel Johnson and The Age of Travel.* Athens: The University of Georgia Press.

Gibson, J. (1887). *Great Waterfalls, Cataracts and Geysers.* London: T. Nelson and Son.

Gilpin, W. (1768). *An Essay on Prints: Containing Remarks Upon the Principles of Picturesque Beauty, the Different Kinds of Prints and the Characters of the Most Noted Masters, Illustrated by Criticisms Upon Particular Pieces to Which are Added Some Cautions That May be Useful in Collecting.* London: Printed for J. Robson.

Holley, G. W. (1883). *The Falls of Niagara with Supplementary Chapters on the Other Famous Cataracts of the World.* New York: A.C. Armstrong & Son.

Hudson, B. J. (2000). The Experience of Waterfalls. *Australian Geographical Studies*, 38(1), 71–84.

Hudson, B. J. (2001). Wild Ways and Paths of Pleasure: Access to British waterfalls, 1500–2000. *Landscape Research*, 26(4), 285–302.

Hudson, B. J. (2002). Best after Rain: Waterfall Discharge and the Tourist Experience. *Tourism Geographies*, 4(4), 440–456.

Hudson, B. J. (2012). *Waterfall: Nature and Culture.* London: Reaktion Books.

Johnson, S. (1944 [1924]). *Johnson's Journey to the Western Islands of Scotland and Boswell's Journal of a Tour to the Hebrides with Samuel Johnson, L.L.D.* (Edited by R.W.Chapman). London, New York, Toronto: Oxford University Press.

Johnson, S. (1816). Letter to Mr. Robert Levet, August 16, 1774, in R. D. Duppa (ed.), *A Diary of a Journey into North Wales in the Year 1774 by Samuel Johnson L.L.D.* London: Printed for Robert Jennings by James Moyes.

Martin, M. (1703). *A Description of the Western Islands of Scotland.* London: Printed for Andrew Bell.

Mayhew, R. J. (2004). *Landscape, Literature and English Religious Culture, 1660–1800: Samuel Johnson and Languages of Natural Description.* New York: Palgrave Macmillan.

McIntosh, C. (1995). Rhetoric and Runts: Boswell's Artistry', in I. S. Lustig (ed.), *Boswell: Citizen of the World, Man of Letters* (pp. 138–157). Lexington: The University Press of Kentucky.

Pennant, T. (1771). *A Tour in Scotland.* Chester: Printed by John Monk.

Perreten, P. F. (1995). Boswell's Response to the European landscape, in I. S. Lustig (ed.), *Boswell; Citizen of the World, Man of Letters* (pp. 37–63). Lexington: The University Press of Kentucky.

Rashleigh, E. C. (1935). *Among the Waterfalls of the World.* London: Jarrold.

Sharma, S. (1996). *Landscape and Memory.* New York: Vintage Books.

Stott, L. (1987). *The Waterfalls of Scotland Worth Gaun a Mile to See.* Aberdeen: Aberdeen University Press.

Chapter 4

The Art of the Railway Poster: Travel, Tourism and the Visual Representation of Place in Britain, 1920–1955

Stephen Williams

Introduction

In industrial nations such as Great Britain, the relationship between railways and tourism development has been influential and intimate. By bringing both established and emerging touristic places within the range of major urban industrial populations and enabling organised, increasingly affordable and efficient mass travel, railways contributed significantly to the development of a wide range of places as tourist destinations: seaside resorts; inland spas; historic towns; popular beauty spots and the wilder and remoter countryside (Urry, 1990; Towner, 1996; Walton, 2000; Williams, 2009). For the last quarter of the nineteenth century and for much of the first half of the twentieth century, railways provided the primary means of tourist travel in Britain.

However, whilst the physical and structural influences that railways exerted on the development of tourism are generally understood, their role in influencing public perception and understanding of space and place through the media of their publicity has been accorded less attention than perhaps it merits. As the epitome of modernity, the railways transformed social space and social relations through the compression of time and space; created new systems of social and physical order; regularised and fixed rituals and patterns of consumption; and, in their construction, demonstrated new levels of human control over nature (Hewitt, 1995; Carter, 2001). But as this chapter will aim to demonstrate, through the media of their publicity, railways also played an important role in shaping how places were represented, imagined and, ultimately, understood by the travelling public. This process was particularly important in the arena of leisure travel and tourism that tended to dominate railway advertising in the period between the two world wars.

The chapter is arranged in two parts. In the first, *The Art of the Railway Poster*, the development of the poster is considered and its influence explained in relation to the use of semiotics and the capacity to act as an instrument of power to transmit ideologies and construct place identity through the tourist gaze. In the second part, *Artistic Imagination and Tourist Places*, several discourses of tourism evident within railway posters of the period are identified and discussed. These discourses are distilled from a content analysis of a sample of some 400 posters from the archive of the Science and Society Picture Library. Each poster was considered in relation to five key variables: the landscape or setting; the presence or absence of people; the characteristics of people (as reflected in age and gender, dress codes or styles, and the composition of groups); activity; and, more subjectively, the prevailing mood or moods that the image sought to convey.

The Art of the Railway Poster

As commercial companies operating in a highly competitive market, railways were active promoters of their services and the role of advertising was central to that process. Initially this had been achieved through the use of printed handbills and notices that utilised text alone, but with the development of low-cost, colour lithography from about 1890 (which paved the way for mass production of cheap, printed images), railway advertising began to adopt the use of poster designs in which images, rather than text, became dominant (Shackleton, 1976; Palin, 1987; Cole and Durack, 1990). At first the new style of poster tended to employ small images that were usually taken from printer's stock and set within over-elaborate and often poorly-conceived designs. However, after about 1910, more creative approaches began to emerge that placed graphic and fine art at the centre of railway poster design and which consequently shaped an important new relationship between art, as a means of place promotion, and the development of travel and tourism.

In 1923 the government approved a restructuring of the railway industry that grouped the many small companies that had previously existed into just four new companies. Central to the new approach to publicity was the conscious decision by the new companies to employ or commission some of the finest artists and graphic designers of the day to produce poster art. Their collective output endowed railway stations with striking, evocative and highly influential designs that not only shaped travel patterns but also constituted works of popular art, some of which were displayed in galleries and collected as prints by an enthusiastic public. Leading exponents of the new art of the railway poster included Norman Wilkinson R. A. (1878–1971) who worked extensively for the London, Midland and Scottish Railway (LMS) and Tom Purvis (1888–1957), who was one of several commercial artists retained by the London and North Eastern Railway (LNER). Wilkinson had produced ground-breaking designs for the London and North Western Railway before the First World War and in 1924 persuaded the LMS to commission 16 members of the Royal Academy (including Maurice Greiffenhagen (1862–1931), Sir William Orpen (1878–1931) and Stanhope Forbes (1857–1947)) to produce a set of original paintings for a new promotional campaign that, in the words of a contemporary, 'placed art on the hoardings' (Cole and Durack, 1992, p. 12). Purvis, a graphic artist of great skill, produced a series of highly original designs for the LNER in the 1930s (see for example Figure 4.1, a poster designed by Tom Purvis for the LNER. This was one of a series of six posters that could be displayed either separately, or linked together as a composite illustration showing typical

Figure 4.1 East coast joys, 1931

Source: Reproduced with permission from the Science and Society Picture Library.

tourist activity on the North Sea coastline). His posters were strongly influenced by the art deco movement and provided images of great visual impact, several of which now reside in the collections of major international galleries including the Museum of Modern Art in New York.

Although railway posters might be designed to promote any aspect of the companies' services, the links with leisure travel were especially strong. The blossoming of poster art after 1920 coincided with a significant growth in the market for holiday travel upon which the railways sought to capitalise (see Walvin, 1978; Towner, 1996; Walton, 2000). In this context, the poster became highly influential on the nature and direction of tourism development in the period between 1920 and 1955, in both a literal and a figurative sense. Through their advertisements, posters and travel guides, the railway companies aimed to harness emergent socio-economic trends (including rising levels of prosperity; increased leisure time; and increases in the incidence of holidays with pay which became a legal entitlement in Britain in 1938), to reinforce existing tourism markets and to shape new ones.

How was this achieved? Critical analysis of processes by which poster art influenced leisure travel suggests four key ideas. First, the railway poster may be considered as a primary mechanism through which dominant ideologies were transmitted and identities, not least of place, were constructed and confirmed. Hewitt notes the powerful ideology of posters as 'encoding notions of orderly pleasure and social power within an economic structure where leisure was being systematically commodified and within social structures where such pleasure was determined by class relations' (1995, p. 307). Critically, the gaze that most railway posters constructed was a middle class, bourgeois gaze. This reflected the fact that at this time the middle classes remained the primary market for holidays, as distinct from the day trip market that was still predominantly working-class in character. The gaze that posters offered had to conform to the social and gendered position of the intended observer, while at the same time confirming it. Consequently, conscious decisions were taken to produce posters with high artistic merit that would resonate with middle class people of taste and provide a welcome antidote to the rising tide of tasteless advertising that was tending to blight both urban and rural landscapes of inter-war Britain (Hewitt, 2000).

Bourdieu's (1984) arguments around the exercise of taste and its relation to social distinction through patterns of conspicuous consumption are highly relevant here. In inter-war Britain, leisure travel was already one of the social arenas in which new class distinctions were played out, drawing on an established tendency for touristic places, especially resorts, to adopt contrasting social tones (see Soane, 1993; Walton, 1994). As the illustrations in this chapter suggest, railway posters, in their contrasting styles, clearly codified different places for different types of experience: the noisy, fun-filled resorts; the more sedate, genteel coastal places; the remoter countryside of rugged coastlines or wild mountains. In so doing, poster art sometimes simply reinforced existing identities of place, but it also sought to transform public appreciation of alternative places as potential destinations for leisure travel. Posters were, after all, advertising destinations that the railways were encouraging people to visit, for the companies' commercial gain.

These transformations in the popular understanding of place were achieved, secondly, through the use of the poster as an instrument of power. Foucault (1972) interprets power as the ability to structure the possible fields of actions of others and this ability is based within his concept of 'discourses' (specialist bodies of knowledge that assume positions of influence in society because they are felt to be more reliable than others). It can be argued that in the inter-war period in Britain, the railway companies occupied such a position. As 'experts' in the business of travel, operating in societies that were coming to terms with leisure travel as an embedded feature of modern life, the discourses of place that were developed by the railways through their poster advertising must have been interpreted by many as intrinsically reliable and, consequently, influential. Of course, as with any marketing device, the consumer develops a reflexive response to the messages offered in the poster, perhaps accepting some elements whilst choosing to disbelieve or be ambivalent towards others. Yet there remains a tendency for any narrative, if repeated sufficiently often and from a position of perceived authority, eventually to become an uncontested 'truth' (Thomas, 1997).

Thirdly, it is evident that railway posters reinforced the impact of their messages to potential holiday travellers by adopting an essentially *touristic* way of seeing. This may have been an instinctive response rather than a planned strategy, but its presence is undeniable. Through the specific circumstances of their production as commissioned art, the posters predominantly reflected an outsider's view of place that was shaped by a complex blend of the real with the imagined (Thomas, 1997). This approach works because it is entirely consistent with the imaginative constructions of place that tourists, as outsiders, also acquire and which form an essential component in the tourist experience. As a part of that experience, initial perceptions that may have been shaped by the poster, what Crang terms 'a structure of expectation' (1997, p. 361), would be validated or modified through the act of travel and, particularly, through sightseeing. Consequently, part of the craft of the poster was that, subconsciously, it generally positioned the viewer as a sightseer, (for example, immersed in the activity of the resort or as a detached observer of landscape or seascapes) and in its design, exhibited the same selective and discriminatory tendencies that shape practices of sightseeing (Thomas, 1997). Here there are parallels with modern tourist practices of photography. Photographs were seldom deployed in railway posters of this period, but many of the designs were shaped by a 'photographic eye' that framed settings from known vantage points and alighted on features of the place that could be guaranteed to attract the visitor.

In shaping these partial and selective readings of different places as tourist spaces, poster art was highly dependent upon the use of semiotics to embed meanings and values to achieve the intended effect. The science of semiotics has been widely established within a range of academic literatures and need not be repeated here in detail (see, *inter alia*, Eco, 1976; Bignall, 2002; Crow, 2003;

Chandler, 2007). The essential consideration is that any sign (in this instance the poster) functions through a relationship between *signifiers* that seek to represent the object (the visual content of the poster) and the *signified*, which is the object itself (the places and activities that are represented in the poster). To work at any level, the semiosis of the poster needs to draw on a background understanding that relies on recognisable, figurative conventions and competencies that are shared by both the producers of the images and the viewers. This shared understanding enables the viewer to make sense of the poster (Watts, 2004). However, once an initial understanding is established, the poster may then draw on wider frames of knowledge to develop not just literal meaning (denotation), but also metaphorical meaning (connotation). Hence, for example, the myriad of inter-war railway posters that promoted the seaside denoted the coast through the use of identifiable images of sand, sea, headlands and harbours, but often connoted more complex notions around escape, rest, relaxation, fun or, in particular situations, glamour or the exotic, as essential but largely unseen attributes of the places in question.

Central to the argument in this chapter is that whilst the figurative representation of tourist places stimulates initial interest and exerts some level of attraction to potential travellers, it is the connotations embedded within the posters that are ultimately more influential on tourist choices. In the age of the railway, posters would have been particularly influential on how travellers would have understood the destinations that the posters encouraged them to visit. Most inter-war posters are images that contain little factual information; their meanings and messages are therefore encoded through their art. Hewitt (2000) argues that our relationship to a poster is conditioned by the setting in which it is viewed. If the main image of a poster is placed in a gallery we 'read' it as art, whereas when it appears on a hoarding we are drawn instinctively to the question of what it is selling. Since we assume it is selling something, the messages need not be too direct, provided they are equally not too puzzling or oblique. Railway posters were normally perused as people waited for trains, so their artistic qualities and the embedded messages could be considered and this considered response was crucial to their success. In this sense, the 'art' of the poster lay not simply in the creative, artistic or technical skills required by the production of the image, but in the more subliminal articulation of complex ideas and values.

Artistic Imagination and Tourism Places

The artistic imagination that is evidenced in railway posters spanned a wide variety of themes. These included an important body of work that captured the spectacle of railway engineering and sometimes the working lives of the men and women who ran the railways, the latter seeming almost to anticipate the post-modern tourist fascination with the working lives of others. However, the content analysis of posters that are overtly related to leisure travel suggest five, distinctive discourses of tourism and place. (Note: This framework excludes the fundamental association between leisure travel and pleasure or, in some contexts fun, which is virtually ubiquitous and thus provides no meaningful basis for distinguishing how places were represented).

Health and Well-being

Unsurprisingly, perhaps, with the exception of the pleasure motive, the most pervasive discourse in railway poster art promoted associations between tourism, health and well-being. In part this reflected an established tradition that dated from the early development of spas and, especially, seaside resorts, which had drawn on beliefs in the restorative qualities of mineral or sea waters and sea air to underpin their growth. However, by the 1920s, a more broadly-based belief in the redemptive properties of sunlight, fresh air and encounters with nature had also begun to capture the popular imagination (Morris, 2009).

Poster art mirrors this relationship in several ways. Some are overt, such as references to the climatic advantage of particular places. Both the Great Western Railway (GWR) and Southern Railway (SR) were relentless in emphasising the sunshine of the southern and western areas of Britain that they served, whilst the LNER, confronted with the more awkward challenge of promoting the colder east coast and the North Sea, alighted on the presumed health benefits of the invigorating qualities of sea air. In poster art this association was first proposed as early as 1908 in John Hassall's[1] famous evocation of the Jolly Fisherman skipping along the 'bracing' sands of Skegness, but the LNER continued to advocate the climatic benefits of the cooler east coast and, particularly, its lower rainfall, by adopting the epithet of the 'drier side' to describe many of the tourist places it promoted.

Although references to local climate were common in railway posters, the discourse of health and well-being was more widely, although subtly, encoded through the appearance of the people shown in posters and the activities in which they were engaged. In contrast to the realities captured in contemporary photographs of inter-war tourists, the population of railway posters were universally shown as fit, young, attractive and, reflecting the growing fashion for sun-bathing, frequently as bronzed (Morris, 2009). People are presented in relaxing situations on the beaches or promenades (to emphasise the restorative qualities attached to tourism and many tourist places), or engaged in wholesome activity, such as walking, swimming, boating or, especially, golf, for which there was a significant niche market for railway travel in the 1930s.

Purvis's poster for the LNER shown in Figure 4.1, which was one of a series of six, linked designs issued in 1931, captures this discourse to perfection. The brief text

1 John Hassall (1868–1948) was an established illustrator and cartoonist who became one of the most influential poster artists of the early twentieth century. A member of the prestigious London Sketch Club and a founder of his own art school, his work as an illustrator was widely known. The image he created of the 'Jolly Fisherman' has continued to be used to promote the resort of Skegness to the present day.

makes the essential reference to the dryness of the east coast climate whilst, through its use of the word 'joy', offers a more subtle prospect of enhancement to the mental state of the traveller. Through the disposition of the figures and the calm sea, the image is infused with a profound sense of relaxation, yet their activity as swimmers, albeit caught in repose, connects directly to notions of health through activity.

Landscape, Nature and Escape

The sense of personal well-being that pervades Purvis's brilliant design links well to the significant body of poster art based on representations of landscape and which connoted notions of escape and the celebration of nature in settings such as picturesque coastlines, lakes and woodland, or remoter mountain regions. These constructions of tourist space, like the discourses of health and well-being, relied for their effect upon value judgments that were quite well established and which drew upon a more extended process in representing both the sea and the wilder environments of mountains and moor as places to be enjoyed by tourists. In the nineteenth century, in particular, the taste for the picturesque; the new sensibilities of the Romantic Movement; the sublime appreciation of natural spectacle; and the rise of new natural sciences such as geology and natural history, had begun to capture the tourist imagination (see, *inter alia*: Andrews, 1989; Bunce, 1994; Urry, 1994; Corbin, 1995; Williams, 2003). Consequently, wild environments that were once shunned as dangerous, uncivilised, or simply unappealing were re-evaluated as places of inspiration to be visited and enjoyed. By the 1930s, railway posters were clearly trading on a public appetite for a sense of escape, of solitude and quietude that contrasted directly with the noisy, grimy, urban settings in which the majority of people resided.

Part of that appetite for rural tourism found a new outlet in the outdoor movement, through which activities such as walking, climbing, cycling, camping and youth hostelling were becoming extremely popular (Walker, 1985) and which were reinforced by railway advertising. In many of the posters that were centred on landscape and nature, the invitation to active exploration, whilst present, is often indirect, but some railway companies used poster art to promote active enjoyment of the countryside and its scenery. The Southern Railway, in particular, ran a successful poster campaign in the 1930s advertising conducted country walks and deploying picturesque images of traditional farming scenes and wildlife to foster a particular set of perceptions on the part of potential customers.

Posters that promoted the appeal of landscapes and nature tended to deploy a number of common devices to encode their message. Importantly, in selling these places to tourists who, in many situations would be new or relatively inexperienced travellers to the places that the posters depicted, the destination had to appear to be 'accessible', not just in terms of the ability to travel but rather, the wider ability of the visitor to cope with the place on arrival. Consequently, posters seldom presented mountain or seascapes as unduly wild or untamed. With occasional exceptions the weather is depicted as calm or benign and although lighting may be dramatic, perhaps with sun behind high, cumulus clouds or as shafts that illuminate the landscape selectively, there is little sense of threat. Exaggeration of relief was sometimes deployed to heighten the dramatic qualities of the scene and, very significantly, with the occasional inclusion of figures such as shepherds (added for picturesque effect or to heighten the sense of an encounter with a different world), these posters are generally devoid of figures. Through this device, the images conveyed the essential sense of escape and solitude that was clearly important to how tourists were encouraged to understand these places. Importantly, though, escape or solitude should not be confused with isolation (with all its attendant risks), so signs of settlement or human occupation are common elements in these designs too. The isolated cottage or a foreground of cultivated lowland to which mountains form a more distant backdrop, provided the viewer with that critical sense of reassurance.

Figure 4.2, a poster designed by Charles Baker for the LMS, typified a romanticised promotion of remoter rural areas that was widely practised in travel literature in the inter-war years. It provides a (literally) beautiful example of this approach. Produced in 1933, it shows a stylised view of Snowdonia (which was a destination area the railway was keen to promote) that contains most of the essential visual elements. The mountain, Tryfan, provides a dramatic backdrop that draws the

Figure 4.2 Snowdonia, 1933

Source: Reproduced with permission from the Science and Society Picture Library.

gaze, whether as sightseer, walker, climber or, indeed, artist, and is clearly intended to fire the adventurous imagination. The motes of sunlight emphasise the natural qualities of the scene, but excepting the single farmstead that signifies people are still at hand, it is an empty landscape that invites solitary exploration.

Exclusive Pleasures

Although Baker's poster of Snowdonia offers an implicit invitation to escape to a romantic and remote landscape, there is, within the sub-text, a sense in which tourist places of this type were also constructed as exclusive. Matless (1998) argues that whilst the promotion of enjoyment of landscape and nature between the two world wars was outwardly a process of democratisation, such activities were often encircled by codes of practice and moral behaviours that were exclusive in their effect. These expectations presumed a vulgar 'other', the tourist who lacked the sensibilities to appreciate the subtleties of landscape and adopt the reflective, contemplative gaze required to appreciate fully these places and derive the restorative benefits that an encounter with landscape and nature was expected to deliver. Similar exclusions are evident in the many hundreds of posters that promoted historic sites. Posters in this genre routinely focused on great buildings, particularly cathedrals and castles; places of historic significance; or the lives of key people from history. But the unstated implication encoded in these images is that only the educated and appreciative visitor need travel to these destinations, the connoisseur perhaps; someone who can be uplifted by the architecture of a great cathedral or the life of an historic figure, or appreciate the vernacular buildings of a medieval town.

The exclusivity of landscape or historic places that is implied in many of these posters is subtle, in contrast, posters advertising resorts are much more overt in proposing relationships between places and particular styles of tourism and, especially, tourists. The exclusivity of places like Snowdonia is signalled by an *absence* of people in the scene; in contrast the exclusivity of resorts is signalled by the *presence* of people who, by their appearance, indicate the social tone to which the resort aspired.

The representation of people in railway posters is instructive. The elderly are uniformly excluded (except when presented as local 'characters') and, except in the promotion of the most popular lower-class resorts, so too are the working classes, large family groups and the noisy and unruly excursionists. Instead posters are generally populated by young couples, by small groups of young people of both mixed and single sex, and sometimes by small family groups. The people are fit and physically attractive; smartly turned out in the latest fashions and engaged in 'appropriate' activity such as promenading, healthy sports and games; or arranged in relaxed social groups. The gaze that is represented in these posters is self-evidently middle class and, in Palin's words 'outrageously selective' (Palin, 1987, p. 6) and whilst these stylish tourists may be cast as observers of more boisterous activity that is evident in the background, the message within these posters is that exclusivity and class distinctions are clearly maintained in the places in question.

The presentation in poster art of tourists as detached observers of places is a recurrent theme. This design, as seen in Figure 4.3, a poster produced by Austin Cooper for the LNER, is one of several that used this device to promote the east coast resort of Scarborough. This poster presents a typical example, in which a young, affluent, couple, fashionably dressed and clearly at ease, recline on a vantage point overlooking the resort of Scarborough. Significantly they are placed apart from others so that no one else invades the exclusivity of the space they have found. The resort itself is represented only as a background scene from which the two figures are physically detached and take on the role of observers rather than participants in the activity of the place. This visual disengagement of the tourist who is positioned as an observer of the place is a recurring motif in some strands of railway advertising, particular in resorts that aimed to maintain a high social tone by highlighting the class of people that it was intended to attract. As a consequence, such images offer a surprisingly imprecise sense of place, tending instead to emphasise intangible attributes of exclusivity rather than any tangible amenities the resort may have possessed.

Figure 4.3 Scarborough, 1932

Source: Reproduced with permission from the Science and Society Picture Library.

Encountering Others

Although some railway advertising suggested the exclusivity of particular tourist spaces, a more significant theme is the encounter with others. These encounters are represented in a variety of forms, however three themes predominate: encounters with the exotic; with the erotic; and with tradition.

Although tourism has become an embedded feature of daily life in the post-industrial era (Franklin, 2003), in the inter-war years popular understanding of leisure travel would have instinctively set the activity apart, bounded in both time and space and imbued with a special significance in the annual cycle of events. Consequently, resorts and holiday regions often acquired an exotic quality that poster advertising reinforced as an opportunity to encounter places of difference. Posters placed bright images of dreamlike landscapes or vibrant seaside resorts, in which that sense of the exotic was enhanced by the use of vivid colour and strong light, against an often grim, noisy and polluted backdrop of industrial places, including, of course, the railway station itself. The parallel of the poster and the setting established an essential contrast between the everyday and the exotic that was entered, at least initially, through the imagination and then subsequently consumed through travel to the destination.

Today the exotic is associated mainly with foreign places, but it is important to appreciate that in the emerging tourism markets of the middle twentieth century, many parts of Britain would have appeared foreign, and hence exotic, to sections of the population. Thomas (1997) coins the concept of the 'domestic exotic' in his reading of Cornwall as presented in posters, an ambivalent encounter that blends the strange with the familiar and which accounts for the subtle sense of 'foreignness' that tourists experience in some transitional locations such as the far west. The promotion of Cornwall by the GWR, whilst often evoking a sense of the foreign through its references to a 'Cornish Riviera', simultaneously refers to an embedded sense of English identity. The latter is signified through the solidity and sense of traditional community values that are presented repeatedly in images of Cornish coastal villages, with their fishing fleets and their picturesque populations of fisherman and local people, who inhabit a landscape of vernacular architectural styles (see Figure 4.4).

The promotion of Cornwall by the GWR relied heavily on popular perceptions of the west as a repository of 'old world' traditions and practices. A version of the Come to Old World Cornwall poster (Figure 4.4), designed by S.I. Veale, was used in the USA by Cunard Line to attract American tourists to the UK. It offers a quiet and contemplative view of one form of seaside tourism, but many resorts were constructed through their advertising as places of excitement, activity, fun and contact with others. Here we encounter the erotic.

Resorts and, particularly, their beaches, have come to be seen as liminal spaces (Shields, 1990; Preston Whyte, 2004), 'places of transition that lie on cultural margins, places of ambiguity in which normal social codes and practices are suspended and which permit alternative codes and behaviours to surface' (Williams, 2009, p. 268).

Figure 4.4 Come to old world Cornwall, 1931

Source: Reproduced with permission from the Science and Society Picture Library.

The carnivalesque[2] nature of many popular resorts offered endless possibilities for encounters with strangers as part of boisterous crowds and incidental social contact became an expected part of the pleasure of visiting resorts. When these encounters became contextualised within the forms of inverted behaviour that Graburn (1983) suggests typifies modern tourist practices, such as relaxed dress codes, they contributed directly to the highly sexualised atmosphere that pervaded popular resorts, especially amongst younger adults (see, for example, Cross (1990) on Blackpool in the 1930s). Beach wear, in particular, became progressively more revealing during the inter-war years as the practices of sun bathing and recreational swimming became more popular, and through exposure of large areas of the body to the gaze of strangers, such incidence of public semi-nudity inevitably eroticised the beach as a social space.

Railway advertising was not slow to make the connection. After 1920, resort posters commonly used the figure of the attractive young woman, clad only in swimwear and normally looking towards the viewer with a waved greeting. The image reinforces notions of the seaside as a site of romance or, more simply, of potential sexual encounters. There is, for example, the suggestion of romance in Figure 4.3, but in Figure 4.5, William Barribal's design for the LNER which captures the sense of carnival to perfection, the sexualisation of space is obvious. This is a particularly interesting example of the presentation of a resort as an eroticised space because the date of its production (1925) is comparatively early for an image in which there are few subtleties. Although there is a mix of people and activity in the background, the fore- and middle-ground is dominated by young, attractive women. Their noisy, waved greetings and the urgency with which they appear to be advancing towards the viewer, whom the poster clearly constructs as male, the seductive glance of the female swimmer in the water and the graphic depiction of several female bodies, compose an image that, by the standards of the day, would be considered highly sensual. Only the male swimmer and the dog strike an incongruous note in an image that fizzes with erotic possibilities.

Nostalgic Modernity

One of the key encounters with others that railway posters also offered was the chance to engage, albeit perhaps fleetingly and in a valedictory way, with a traditional Britain that was, by the 1930s, disappearing rapidly. The phrase 'nostalgic modernity' attempts to capture the tensions within this process. On the one hand, the

2 The concept of the 'carnivalesque' originates in the work of Bakhtin (1984). For Bakhtin, carnival is a spectacle in which all participants are actors and in which normal hierarchical social positions are suspended as participants are liberated from norms of etiquette and decency that are imposed at other times. Shields argues that 'carnival produces a momentary social space based on the politics of pleasure and the physical senses' (1990, p. 95). In Shields's work, the carnivalesque is closely associated with the liminality of resorts and beaches, where the liberated forms of behaviour associated with the carnivalesque were widely accepted and, as a consequence, flourished.

Figure 4.5 Bridlington, 1925

Source: Reproduced with permission from the Science and Society Picture Library.

rising tide of modernity (as reflected in increased mobility, new technologies, new media and new forms of communication) served to raise awareness and enable wider personal engagement with traditions and the past. Yet, at the same time, modernity was contributing directly to the consignment of that past into memory. Matless, for example, comments that the new taste for leisure motoring encouraged a 'modern practice in pursuit of an older England … the petrol engine allowed a nostalgic passage to an old country, its landscape and its rustic inhabitants' (1998, p. 64). However, this new taste was not confined to motoring; the steam locomotive allowed similar access to a nostalgic England too, albeit within a similarly ambivalent relationship. As Watts explains, the railways embraced and shaped the modern world, but through their advertising often constructed an idealised, rural counterpart to it, a place of 'refuge and spiritual renewal' (2004, p. 35).

The construction of this nostalgic tourist gaze was part of a wider process of introspection that has been detected in the cultural 'climate' of inter-war Britain, and England in particular. Light, in exploring literary representations of English national identity from feminist perspectives, suggests that the aftermath of the 1914–18 war saw 'a move away from a formerly heroic and officially masculine public rhetoric of national identity … to an Englishness at once less imperial and more inward-looking, more domestic and more private' (1991, p. 8). Similarly, Watts (2004) coins the concept of a 'deep England', a personalised immersion, through travel, in the beauties of an unspoilt countryside and drawing on the tradition of a romantic gaze that emphasised solitude, privacy and a semi-spiritual relationship with its object.

These shifts in understanding were reflected in a range of contemporary media. Fiction, music, film, art, advertising, travel guides and railway posters became widely imbued with a collective, wistful, celebration of traditional landscapes, communities and their associated aesthetic, social and moral values (Gruffudd, 1994). Rural landscapes, in particular, were widely constructed as a repository of authentic values and an expression of true national identity, so much so that inter-war poster art was easily transformed into wartime propaganda that aimed not just at recruitment to armed forces, but also at galvanizing a collective, national desire to protect these essential attributes.

Figure 4.6, designed by Adrian Allinson for the Southern Railway, typifies this approach. This nostalgic view of a British countryside that was quickly receding into popular memory epitomised a representational style that was used extensively in a new generation of motoring guides as well as railway posters. This is an image that captures a harmonic view of the countryside, an aesthetically pleasing blend

Figure 4.6 Isle of Wight, 1946

Source: Reproduced with permission from the Science and Society Picture Library.

of geology, topography and scenery that many in the inter-war preservationist movement advocated as desirable. The farmstead is enfolded by a domesticated, rolling, landscape of woodland and cultivated fields, in which farm workers load a hay wagon with their harvest; a scene that is pervaded with an over-riding sense of peaceful tradition and continuity. This poster was produced in 1946, emphasising, perhaps, the power of the nostalgic gaze and its ability to summon a vision of a place that the exigencies of war had probably damaged irretrievably but which, if the poster was to be believed, could still be located if one knew where to look.

Equally evocative, and no doubt as influential, were the many posters that drew on heritage themes to promote tourist places. The nostalgic gaze is principally about individual and collective identity and whilst it adheres to landscape as one expression of identity, history is equally influential in this regard. There is a tendency to consider heritage tourism as a very contemporary, almost post-modern creation, but railway advertising of the 1930s shows quite clearly that heritage was very much a part of defining the nature of some places as tourist space. Posters appealed to patriotic sensibilities through traditional scenes accompanied by captions such as 'This England of Ours', in ways that would have been clearly understood in an era of rising international tension and the twin challenges of communism and fascism.

Conclusion

Richards and MacKenzie characterised the age of the railway in terms of 'the slow, sure and steady progress of social improvement … an age of hope, of optimistic belief in the future, unashamed aspiration for better days and better conditions in the world' (1986, p. 17). These are sentiments that are precisely captured and conveyed in railway posters of the inter-war and early post-1945 period.

This chapter has argued that as a form of popular art, railway posters of the period between 1920 and 1955 were an important influence on how tourism places were represented and understood. With some small exceptions, railways neither invented the rituals of tourism, nor, in isolation, did they create tourism places. But through the use of art in their advertising they certainly reinforced place identities and their rituals and, in an era of rapid expansion in domestic tourism in the UK and at a time when the majority of tourists still travelled by train, the influence that posters exerted in promoting particular tourist places and associated styles of tourism should not be neglected.

In retrospect it is evident that the middle decades of the twentieth century represent a key transition period between emergent phases of tourism before 1914 (in which holiday travel was still a largely exclusive domain), to the creation of genuine mass tourism in the decades after 1960. The discourses of tourism that are revealed in poster art of the period reflect that transition with great clarity, drawing, as we have seen, on established associations such as health and well-being or the celebration of nature and landscape, but also reflecting contemporary change in areas such as social relations or the quest for the exotic, and the rising tide of nostalgia that accompanied rising international tensions and the onward progress of modernism, especially after 1945. Railway posters performed a dual role, on the one hand reflecting wider processes of social and economic change, but in so doing, also helping to drive that change forward through the influential nature of their publicity.

There is, of course, reflexivity in this process. Light observes that 'the inter-war period was one in which the English began to see images of themselves and their cultural behaviour … at the pictures and in magazine photography, as never before' (1991, p. 215). The railway poster was clearly a part of that same process of cultural reproduction, imagining tourist space based partly on what it already was, but conveying to the millions of new tourists that visited these places as rail travellers between 1920 and 1955, a sense and an understanding of what it could, or should be.

Acknowledgements

I am grateful to friends and colleagues at Staffordshire University: Paul Barratt, Fiona Tweed and Allan Watson, for their comments on a draft of this chapter. The images presented in the chapter are reproduced by courtesy of the NRM/Science and Society Picture Library.

References

Andrews, M. (1989). *The Search for the Picturesque: Landscape Aesthetics and Tourism in Britain 1760–1800*. Aldershot: Scolar Press.

Bakhtin, M. M. (1984). *Rabelais and His World*. Bloomington: Indiana University Press.

Bignall, J. (2002). *Media Semiotics: An Introduction*. Manchester: Manchester University Press.

Bourdieu, P. (1986). *Distinctions: A Social Critique of the Judgement of Taste*. London: Routledge and Kegan Paul.

Bunce, M. (1994). *The Countryside Ideal: Anglo-American Images of Landscape*. London: Routledge.

Carter, I. (2001). *Railways and Culture in Britain: The Epitome of Modernity* Manchester: Manchester University Press.

Chandler, D. (2007). *Semiotics: The Basics*. London: Routledge.

Cole, B. and Durack, R. (1990). *Happy as a Sand Boy: Early Railway Posters*. London: HMSO.

Cole, B. and Durack, R. (1992). *Railway Posters 1923–1947*. London: Lawrence King.

Corbin, A. (1995). *The Lure of the Sea: The Discovery of the Seaside 1750–1840*. London: Penguin.

Crang, M. (1997). Picturing Practices: Research Through the Tourist Gaze. *Progress in Human Geography*, 21(3), 359–373.

Cross, G. (ed.) (1990). *Worktowners at Blackpool: Mass Observation and Popular Leisure in the 1930s*. London: Routledge.

Crow, D. (2003). *Visible Signs: An Introduction to Semiotics*. Crans-pre-Celigny: AVA Publishing.

Eco, U. (1976). *A Theory of Semiotics*. Bloomington: Indiana University Press.

Foucault, M. (1972). *The Archaeology of Knowledge*. London: Tavistock Publications.

Franklin, A. (2003). *Tourism: An Introduction*. London: Sage.

Graburn, N. (1983). The Anthropology of Tourism. *Annals of Tourism Research*, 10(1), 9–33.

Gruffudd, P. (1994). Selling the Countryside: Representations of Rural Britain, in J. R. Gold and S. Ward (eds.), *Place Promotion: The Use of Publicity and Marketing to Sell Towns and Regions* (pp. 247–263). Chichester: John Wiley.

Hewitt, J. (1995). East Coast Joys: Tom Purvis and the LNER. *Journal of Design History*, 8(4), 291–311.

Hewitt, J. (2000). Posters of Distinction: Art, Advertising and the London, Midland and Scottish Railway. *Design Issues*, 16(1), 16–35.

Light, A. (1991). *Forever England: Femininity, Literature and Conservatism between the Wars*. London: Routledge.

Matless, D. (1998). *Landscape and Englishness*. London: Reaktion.

Morris, N. J. (2009). Naked in Nature: Naturism, Nature and the Senses in Early 20th Century Britain. *Cultural Geographies*, 16(3), 283–308.

Palin, M. (1987). *Happy Holidays: The Golden Age of the Railway Poster*. London: Michael Joseph.

Preston-Whyte, R. (2004). The Beach as a Liminal Space, in A. A. Lew, H. C. M. and A. M. Williams (eds), *A Companion to Tourism Geography* (pp. 349–359). Oxford: Blackwell.

Richards, J. and MacKenzie, J. M. (1986). *The Railway Station: A Social History*. Oxford: Oxford University Press.

Shackleton, J. T. (1976). *The Golden Age of the Railway Poster*. London: New English Library.

Shields, R. (1990). *Places on the Margin: Alternative Geographies of Modernity*. London: Routledge.

Soane, J. V. N. (1993). *Fashionable Resort Regions: Their Evolution and Transformation*. Wallingford: CAB International.

Thomas, C. (1997). See Your Own Country First: The Geography of a Railway Landscape, in E. Westland (ed.), *Cornwall: The Cultural Construction of Place* (pp. 107–128). Penzance: Patten Press.

Towner, J. (1996). *An Historical Geography of Recreation and Tourism in the Western World, 1540–1940*. Chichester: John Wiley.

Urry, J. (1990). *The Tourist Gaze: Leisure and Travel in Contemporary Societies*. London: Sage.

Urry, J. (1994). The Making of the English Lake District, in J. Urry (ed.), *Consuming Places* (pp. 193–210). London: Routledge.
Walker, H. (1985). The Popularisation of the Outdoor Movement, 1900–1940. *British Journal of Sports History*, 2(2), 140–153.
Walton, J. K. (1994). The Remaking of a Popular Resort: Blackpool Tower and the Boom of the 1890s. *The Local Historian*, 24(4), 194–205.
Walton, J. K. (2000). *The British Seaside: Holidays and Resorts in the Twentieth Century*. Manchester: Manchester University Press.
Walvin, J. (1978). *Beside the Seaside: A Social History of the Popular Seaside*. London: Allen Lane.
Watts, D. C. H. (2004). Evaluating British Railway Poster Advertising: The London and North Eastern Railway between the Wars. *Journal of Transport History*, 25(2), 23–56.
Williams, S. (2003). *Tourism and Recreation*. Harlow: Prentice Hall.
Williams, S. (2009). *Tourism Geography: A New Synthesis*. London: Routledge.

Chapter 5

Towards a Realist Ontology of the Holiday: Seduction, Denial and Delusion in the Narratives of Tourism in Twentieth Century Opera

David Botterill

Introduction

In the comic opera *Flight* (Dove, 1998) the character Older Woman is at the airport terminal hoping to meet her new lover, a barman she met on holiday. It is Wednesday. She is clasping a post card received from him containing a post script 'See you Wednesday. Wednesday! Which Wednesday?' she asks herself, and continues 'he said he would be coming, he promised me. My fiancé. Only I haven't heard from him for weeks. I'll be there Wednesday. It's absurd. Holiday romance at fifty-two. Wait till it happens to you'. In the newspaper story titled *Woman Lost Savings to Turkish Waiter* life mimics art:

> A 50 year old-woman has told how she lost her life savings to a 28-year-old Turkish waiter with whom she fell in love on holiday … sold her house and gave £70,000 to … so they could run a bar together. Almost as soon as she transferred the money into his account, her lover disappeared. (*The Daily Telegraph*, 2013, p. 9)

At first sight, operas narratives and newspaper reports might seem strange data sources for a social scientist to use in progressing an ontological analysis of the holiday. For those readers who are familiar with my exploration of critical realist tourism research (Botterill et al., in press; Platenkamp and Botterill, 2013), however, such epistemic relativism will come as no surprise. Neither will the retroductive move from these data to identify objects with powers that enable and constrain what is represented on the surface as a holiday.

Other scholars of tourism have demonstrated the potential of a wide range of representations of tourism in art forms such as in literature, film, popular music, poetry, photography, and fine art for exploring narratives of tourism (e.g. see Crawshaw and Urry, 1997; Francesconi, 2011; Robinson, 2005; Tribe, 2008). In my project, I extend this list of cultural representation to include western classical opera. The insights into narrative tropes that I reported in a meta analysis of twentieth century opera narratives (Botterill, 2012) encouraged me to pursue new lines of analysis within my project, this time based upon the characterisation of the 'tourists' in the operas. These take the form of micro-analysis of the librettos featuring particular arias or exchanges between the characters. To achieve this I selected scenes from three operas; *The Death of Klinghoffer* (Adams, 1991), *Death in Venice* (Britten, 1973), and *Flight* (Dove, 1998), as examples of twentieth century opera where characterisations of tourists take a central position in the musical drama. The tourist characters in these operas are immersed in situations that present what I have called the counter narratives of tourism, that of transgression, inversion and vulnerability, unearthed in my analysis of the narratives of 38 twentieth century operas (see Botterill, 2012). In the opera storyline they are all facing situations where the normal ontological order of a holiday has been overturned, where reality shows an irrational face and where chaos, confusion, ambivalence and tragedy have profound impacts on their personal experiences of being on holiday. I will argue that it is in these contexts that the necessary, but not sufficient, conditions of a holiday, that of seduction, denial and delusion, suddenly and quite shockingly become exposed. For each opera I provide examples of the tourists' narratives found in the libretto and comment upon their translation into the musical score.[1] In the final section, I locate my analysis of narratives of twentieth century opera within the philosophical frame of scientific 'depth' realism to produce an exploration of the social ontology of the holiday.

Selected Opera Scenes

The Death of Klinghoffer

This opera is based on the 1985 Palestinian attack on the cruise ship Achille Lauro. Whilst sailing in the Eastern Mediterranean the vessel was hijacked, and during the event a paralysed Jewish-American passenger, Leon Klinghoffer, was fatally shot and his body thrown overboard (Botterill, 2012). The first selection is taken from Act 2 Scene 1A of the opera. It features a sung conversation between Leon and one of the four terrorists involved in the incident, Rambo, and it is a deeply personalised attack, one on the other, displaying the acrimony felt by these characters of Jewish and Arab identity. I have chosen to use the order of libretto found in a film version of the opera. The film version, directed by Penny Woolcock, and released in 2003 is a dramatised account of the opera and is not a film of an opera production. The order of the original libretto by Alice Goodman has been slightly amended to provide a more compatible libretto for film production. It was shot on location in the Mediterranean on a cruise liner and at various locations in the Middle-East.

The scene provides an entrée into the various levels of meaning contained within the opera. Rambo and Leon Klinghoffer are personifications of the terrorist/

1 Please note that the capitalisation and punctuation of phrases quoted in this chapter have sometimes been altered from the printed opera librettos in order to aid comprehension in written, as opposed to sung, phrasing.

tourist encounter. They represent, on a very personal level, the dramatic clash of oppositional forces that surround one of the most perplexing problems of the twenty-first century, the question of an independent Palestinian state. They also portray the social roles of 'tourist' and 'terrorist'. In this particular scene, the roles are simplified in a rather stereotypical characterisation. Leon is the innocent, ordinary man, somehow accidentally and erroneously caught up in a needless act of violence and terror while on holiday. His character is made more vulnerable by his disability. Rambo is the angry 'irrational' man desperate for revenge and driven to desperate measures to fight against the historical injustice of the loss of his homelands and national identity. And when Leon and Rambo sing their arias, which follow below, the audience can immediately see how the personal dispute between Leon and Rambo that is centred upon their different religious identities is magnified to encompass relations between the English speaking and Arabic speaking regions of the world. Consequently, their differences become secularised and encompass far greater numbers of the world's population and their governments. In this exchange the tourist character loses his naive innocence and the terrorist character displays his anger at Leon and what he represents.

> Leon: We are the kind of people you like to kill. Was it your pal who shot that little girl at the airport in Rome? You would have done the same. There's so much anger in you. And hate, I know how children in the Promised Land learn to sleep underground because of your shelling. Old men at the Wailing Wall get a knife in the back. You laugh. You pour gasoline over women, passengers on the bus to Tel Aviv and burn them alive. You don't give a shit, excuse me about your grandfather's hut, his sheep and his goat, and the land he wore out. You just want to see people die.

> Rambo: You are always complaining of your suffering but wherever poor men are gathered they can find Jews getting fat. You know how to cheat the simple, exploit the virgin, pollute where you have exploited, defame those you cheated, and break your own law with idolatry. America is one big Jew. (to another tourist) … Are you English? Your Balfour Declaration led to the partition and the dissolution of the Palestinian nation. Where English is spoken you will find perversion and all kinds of filth not practised by stealth, late at night but on the street during the day. You wink at sodomy. You laugh at blasphemy. You give no charity to the oppressed.

The scene ends with an emotional reunion between Leon and his wife Marilyn Klinghoffer who are forcibly separated during the day. Unbeknown to Marilyn, Leon had been subject to some rough treatment at the hands of two of his captors and a vitriolic exchange of views. Marilyn and Leon embrace. Leon makes a joke of their situation, and reassuring Marilyn of his own well-being, he sings:

> Leon: This was to be our happy time together. Damn. Oh Damn it M., let's see you smile. Look at that gull, think he'll land in the pool? That's better. We'll

> bring home a tan anyway. When I want to lie down and get out of the sun I'll get a man to wheel me below. One thing less for you to worry about. I should have worn a hat. I should have worn a hat.

It was this overly simplified characterisation of Leon that, in part, fuelled the anti-Semitic controversy that followed the first performances of the opera in the US (Fink, 2005). But it was not just the Leon character on its own. In a scene from the prologue, which was subsequently cut after the first performances of the opera, the Rumour family, neighbours of the Klinghoffers, are depicted in their living room discussing various everyday matters including the Klinghoffer's impending cruise and which is set against the broadcast newscast coverage of the Middle-East crisis. In a published interview with Adams (Christiansen, 2012), he explained that the scene was introduced by Alice Goodman as an element of humour in what is a despairing story. The conversations between Alma, Harry, and Jonathon Rumour were intended to be a satirical addition to the opera. The critics in the first performance received the scene not as intended, a satire on American consumerism, but as an offensive satire on the Jewish way of life. Despite its deletion from the opera this 'set a tone' for a continuing criticism of the way that the Klinghoffers are portrayed as shallow and rather empty tourists particularly when set against the nuanced re-telling of the stories of the lives of the terrorists that are incorporated in three arias sung by Molqui, Mamood and Omar. In his defence, Adams has argued that the choral-like passages of the opera are analogous to the Bach passions and resound as a backcloth against which Klinghoffer's own 'heroic' death is depicted, particularly in the Aria of the Falling Body. Unsurprisingly, Adams resists all accusations of anti-Semiticism in the opera.

In the context of the overall narrative what I take from the above is how determined Leon is to cling to the ideal of the holiday in the face of its obvious disintegration into chaos. This can only be achieved by a supreme act of denial of the appalling, life-threatening, situation he and Marilyn find themselves in. Exactly how fatal the situation will become for him is, of course, not known to them at this time in the drama. But, is the response of Leon to their situation also delusional? The references to the 'tan', the gull landing in the pool, and the absent-minded 'I should have worn a hat' provide the grounds for shifting my analysis from denial of the situation to delusion: a state of mind formulated in the face of an irrational reality. The cruise liner has become a prison where cruelty is handed out on the basis of national identity. Leon and Marilyn are no longer having their happy time together but are forcibly kept separate from each other. They are subject to verbal and physical abuse. They and other hostages are denied access to bathrooms and starved of food and drink. And yet what preoccupies Leon is his thoughtless dependency on Marilyn for everything, if only he had thought about his hat, she might stop worrying about him. It is as if Leon wants to believe that a hat could protect either of them from their captors, a moment of excruciating pathos in tragic circumstances. I will return to the ontological powers of denial, delusion and the ideal of a holiday later in my chapter, but first another example from the second opera in which the power of seduction emerges to join denial and delusion with yet more tragic consequences for a tourist.

Death in Venice

Suffering from writers block and having being encouraged to make a trip to Venice by a mysterious traveller, the author Aschenbach meets Tadzio, a young Polish boy who is residing at the resort with his family. Throughout the opera the opposing characters of Apollo (Tadizo) and Dionysius (variously the wanderer, a gondolier, the hotel manager, a railway port and a hairdresser) compete for Aschenbach's creativity as he struggles with his affections for both Venice and Tadizo. Unable to evade the outbreak of cholera that besieges the city, the author dies in his deckchair gazing out to sea (Botterill, 2012).

I have selected several scenes from *Death in Venice* that I think expose how seduction, denial and delusion interact in the social ontology of Aschenbach's 'holiday'.

Act 1 Scene 1 (Munich): Walking around his home town Aschenbach encounters a traveller, a stranger, (presumed to be a mortal depiction of Dionysus) who urges him to travel South.

> Aschenbach: What is this urge that fills my tired heart, a thirst, a leaping with unrest, a deep desire!
>
> Traveller: Marvels unfold! No boundaries hold you, go, travel to the South. Great poets before you have listened to its voice.
>
> Aschenbach: I have always kept a close watch over my development as a writer, over my behaviour as a man. Should I now, without thought, break my habit, my summer of work in the mountains, to holiday in the warm and lovely South? The break can be justified of course by this growing fatigue, that no one must suspect and that I must not betray in any sign of flagging inspiration. Yes, it can be justified – but the truth is that it has been precipitated by a sudden desire for the unknown. So be it.

This conversation, and Aschenbach's reflections on it, reveals the motives and the anxieties that surround his decision to break with routine and take a holiday. His professional image is dented by the fatigue he feels that cannot be exposed. Images of the 'warm and lovely South' excite his passions but depict, too, a veiled warning of the unknown. We can immediately see that in the character Aschenbach we have a more worldly tourist with a different utopian vision, sensitive to the unpredictability of the events of a holiday, but excited by the exoticism and sensuousness of the south.

Act 1 Scene 3 (The journey to the Lido): Aschenbach has taken a gondola, he sings,

> Aschenbach: Ah Serenissima![2] Where should I come but to you to soothe and revive me. Where but to you, to live that magical life between the sea and the city? What lies in wait for me here, ambiguous Venice, where water is married to stone and passion confuses the sense? Ambiguous Venice.

Then, he realises that the Gondolier (another mortal depiction of Dionysus) has taken a different route and in the exchange that follows Aschenbach displays the anxiety of finding oneself at the mercy of a stranger who decides what they want to do,

> Gondolier: Nobody shall bid me, I go where I choose, I go my own way, I have nothing to lose.

Aschenbach relents realising he can't control what is happening. At the end of the trip he tips the boatman and sings,

> Aschenbach: Mysterious gondola, a different world surrounds you, a timeless, legendary world of dark, lawless errands in the watery night. How black a gondola is – black, coffin black, a vision of death itself and the last silent voyage. Yes, he rowed me well. But he might have done for me, rowed me across the Styx and I should have faded like echoes in the lagoon to nothingness.

Arrived at the hotel and shown to his room by the Hotel Manager (Dionysus again), he sings,

> Aschenbach: So I am led to Venice once again … Now in this beautiful, agreeable place, I intend to give myself to the leisure world for a spell. A pleasant journey, did he say? (referring to a conversation with the hotel manager). The whole experience was odd, unreal, out of normal focus. Was I wrong to come, what is there in store for me here?

Act 1 Scene 6 (The foiled departure). Aschenbach is not enjoying himself. The oppressive heat, the smells of the city, the crowded bustle of the streets and the constant harassment of the street vendors close in around him,

> Aschenbach: Horrible, evil, nauseous. I must go elsewhere, I must find a clearer sky, fresher air.

He leaves the hotel for the train station but there is a mix up with his luggage and he has to return to the hotel,

---

2 Serenissima is a name of the Republic of Venice. The meaning of the word serenissima is the very serene.

> Aschenbach: I become like one of my early heroes, passive in the face of fate. What do I really want? First, I am grief-stricken but must go because of the danger to my health, then I am furious because I am forced to return, but secretly I rejoice. Vacillating, irresolute, absurd.

The first Gondolier is heard in the distance calling for customers. Aschenbach again,

> Aschenbach: The whole experience has been disruptive to my thoughts and to my work; yet in spite of it I feel my spirits rise. Often what is called disruptive is not directed against life, but is invigorating, a renewal …

In Act 2 of the opera, the drama unfolds around rumours and then the eventual confirmation of a cholera epidemic. This meeting with irrationality, with a sinister, untamed, natural element of disease and illness, a threat to human life itself, seems to be a turning point for Aschenbach. He buys a German paper in seeking an authoritative statement on what is happening,

> Aschenbach: Let me see what my countrymen say 'Rumours of cholera in Venice officially denied. Rumours of an incipient plague in Venice officially denied' Ah here it is, 'We doubt the good faith of the Venetian city fathers in their refusal to admit to the case of cholera in the city. German citizens should return as soon as possible'. Ugh! Ugly rumours, rumours. They should be silent. The city's secret growing darker every day, like the secret in my own heart.

The secret he refers to his admission, at the end of Act 1, of falling in love with the young Polish boy Tadzio, a fellow tourist staying at the hotel with his mother and sisters. The family pass by at the moment he sings,

> Aschenbach: They must receive no hint. They must not be told. They must not leave. (and later as he follows Tadzio's family). Ah, Tadzio, Eros, charmer, see I am past all fear, blind to danger, drunken, powerless, sunk in the bliss of madness. Ah!

The seduction is complete, the threat of cholera is denied and in a delusionary response Aschenbach justifies his collusion with the secrets of the state to hide his own, dark, desire.

Act 2 Scene 12 (The lady of the pearls). Aschenbach is provided with several opportunities to talk to the family but he demurs.

> Aschenbach: So I didn't speak! Once again I have failed to make everything decent and above board, missed the opportunity to become myself again, missed the opportunity to regain my reason, my self-possession? What is reason, moral sense? What is art itself, compared to the rewards of chaos? The city's secret,

desperate, disastrous, destroying, is my hope. I will not speak. What if all were dead and only we two left alive?

Of all the operas in my project, *Death in Venice* offers perhaps the most complete demonstration of the complex interplay of powers of seduction, denial and delusion in the social ontology of the holiday. Seduced by the qualities of Venice and by the beauty of a boy and overwhelmed by his passions and senses, Aschenbach is in denial of the ambiguity, disruption, and chaos that surrounds his holiday and, ultimately, he is deluded sufficiently to stay in Venice against all advice and fall victim to cholera.

Flight

Based loosely on the discovery of a refugee who lived for 12 years inside Charles de Gaulle airport, this comic opera set in an airport terminal documents the life of tourists, travellers and airport employees over a 24 hour period. A storm envelopes the airport and causes the characters to interact in unexpected ways. In the opening Act passengers are arriving and join the Refugee and the airport Controller in the airport terminal. This extract from Act 1 Scene 1 captures the opening exchanges between Bill and Tina, a holidaying couple,

Bill: Darling

Bill & Tina: Are you happy? Perfectly! We're here in plenty of time.

Tina: And we've our lucky donkey for luck!

Bill: We don't need luck!

Bill & Tina: Kiss me! We'll feel the way we did before, on holiday.

Bill: Our holiday will be fantastic.

Tina: My costume is just shells and elastic.

Bill: Double fantastic!

Tina: We'll walk on the beach and do things together.

Bill: Like swimming.

Tina: And walking.

Bill: And drinking together.

Tina: And maybe even talking to each other.

Bill: We'll feel the way we did before, on holiday.

Bill & Tina: We won't argue.

Tina: And you won't snore.

Tina: You'll lose a few pounds.

Bill: And you'll lose a few pounds.

Tina: Do you think I need to lose a few pounds?

Bill & Tina: We'll feel the way we did before. Like one person. The whole world disappeared except you … What if we don't feel the way we did before?

Bill and Tina exchange a few words with the Refugee and give him money. The Older Woman enters, 'dressed to kill' (perhaps as a Flamenco dancer), and sings,

Older Woman: Do I look conspicuous? I mean strange? Should I have worn a less flamboyant range? Is it obvious that I may be deranged? I'll just sit here quietly and try to disappear into the general atmosphere. If anyone asks me where I am going, I'll speak French and pretend I'm foreign.

The Older woman isn't going anywhere in fact, as she tells us later in Act 2 Scene 1,

Older Woman: He said he would be coming, he promised me. My fiancé. Only I haven't heard from him for weeks. I'll be there Wednesday. It's absurd. Holiday romance at fifty-two. Wait till it happens to you.

The male and female airline stewards pick-up on her apparent nervousness and wrongly assume that she is anxious about flying,

Steward/ess: Let us escort to your plane.

In response the Older Woman sings,

Older Woman: I'm not going anywhere. I'm waiting for my fiancé. He works in a bar in Majorca. I met my fiancé on holiday. My fiancé! What am I saying? He'll never come. I'm fifty-two and he's young. Why did I come? I should go home … If only he would come. It would be sublime. I'd be reeling. It would be like flying. What am I saying? I'd be reeling. A sublime feeling.

Later the assembled characters are awaiting the announcement of the plane's departure. Older Woman is reading her postcard to another passenger 'Minskwoman' – the pregnant wife of a diplomat en route to a new posting in the Ukraine. Bill and Tina are reading their relationship manual together and the Steward and Stewardess are flirting.

> Older Woman: Hello Sandra. That's me. The weather here is sunny. I have new shoes.
>
> Tina: Page a hundred and three. 'Give your sex life variety'.
>
> Steward: We can sit in the darkness …
>
> Bill: Are you saying you're bored with me?
>
> Steward: … And be romantic!
>
> Older Woman: This is the nice bit. Remember the night, we walked along the beach. It was good.
>
> Steward: You've got nice hair.
>
> Stewardess: I like your shirt. You've identical eyes.
>
> Steward: You've got the thighs.
>
> Bill: It says on page a hundred and four: 'Don't feel obliged to perform, anything you have not done before'.
>
> Tina: What a bore.
>
> Steward: No hurrying, no panic, no stifled cries.
>
> Stewardess: No thirsting or bursting or stupid lies.
>
> Steward: We can just be in this great big green field.
>
> Older Woman: Did I say the sea was blue? In the bar there is a lot to do. Yes, the bar is very busy. PS See you Wednesday. … Wednesday! Which Wednesday?

While the Steward and Stewardess progress with their fantasy and Bill and Tina engage in an increasingly acrimonious conversation, the Older Woman continues,

> Older Woman: He said he would be coming, he promised me. My fiancé. Only I haven't heard from him for weeks. I'll be there Wednesday. It's absurd. Holiday romance at fifty-two. Wait till it happens to you.

At the end of the opera, when all the passengers are finally getting away, she sings,

> Older Woman: I'll have another holiday. Perhaps, I'll meet someone there. In a café or a square. I'm prepared, to travel far.

In these character narratives we can once again see the interplay between seduction, denial and delusion that was also demonstrated by Leon in *The Death of Klinghoffer* and Aschenbach in *Death in Venice*. Bill and Tina are both acutely aware of the state of their relationship and they pour so much hope into the holiday to help them reconnect to past moments. As they fantasise about the sexual possibility of the holiday they enter into denial about the true state of their relationship but here, at least, there is a worrying awareness that the holiday might not 'work' its magic. The Older Woman, on the other hand is out to be seduced by yet another holiday romance and deluded sufficiently to believe her life will be forever different.

Conclusions: Towards the Ontology of the Holiday

From my data, opera librettos, I have argued that seduction, denial, and delusion are necessary but not sufficient conditions for a holiday to exist. I want to make it clear that such claims about the intransitive are always fallible, and historically and culturally contingent, but there is a practical adequacy about them in that they offer a realist explanation of the enduring tendencies of a holiday. In the phrase 'not sufficient conditions' I am, of course, acknowledging many other objects with powers that are also present. Those powers may be material objects such as means of transport or passports or beaches and mountains. They may be exercised and observed, or exercised and unobserved, or simply unexercised, in any given situation. But crucially here I am arguing that social mechanisms exist, that they have powers and that they interact with other objects; natural, material and social, to create what is called a holiday.

Making a retroductive move from the opera scenes, I have proposed three social objects with powers (or generative mechanisms to use critical realist terminology): seduction, denial, and delusion. I want now to make a further retroductive move to ask what makes those powers possible. By making this move I am seeking to dig deeper into the social ontology of a holiday. Here I propose a further generative mechanism, an ideal form of a utopian holiday, as a further necessary but not sufficient condition for a holiday to exist. When Aschenbach sings 'Ah Serenissima! Where should I come but to you To soothe and revive me' and Leon signs to Marilyn 'This was to be our happy time together' and the characters of *Flight* describe aspects of holidays remembered (see below) they begin to sketch out aspects of the utopian holiday. Throughout these opera narratives the powers of a utopian holiday lurk under the surface, but only when chaotic events disrupt the utopian form can we see the full extent of its seductive and delusional power that in many cases lead to a denial of what is really happening on a holiday. It is the utopian ideal of a holiday that seduces, and against it that reality is denied, leading, sometimes, to delusion.

I will end this chapter with a further operatic example of what I am proposing by the mechanism, an ideal form of utopian holiday, and the necessary mechanisms of denial and delusion. At the end of Act 1 of *Flight* the assembled characters anticipate their journeys. They all sing the following aria,

> All: Its beginning. Our turn soon. Our holiday's beginning. Lotions and donkeys. And shells and elastic, and kiddies with buckets, and tans fantastic. Snorkels and sunburn, and towels that are sandy, and novels that make you feel awfully randy. T shirts and tombs, discos, falafels, beers and bikinis, foreigners, camels, pina coladas and olives and feta, and things are just getting better and better. Waves and horizons, the stretch of the day. We always want to be on holiday.

The music works into a crescendo of sound and speed until the final few bars of the act prick the bubble of holiday ideal,

> Controller: Due to circumstances beyond our control, we regret to say that flights will experience some delay. For how long we can't say. Electric storms are coming our way. Please wait patiently till skies clear.
>
> All: Storms! Oh dear.
>
> Refugee: Now just like me you're stuck here. Just like me.

The natural powers of a storm paralyse the characters and in so doing expose the necessary and sufficient conditions for a holiday to exist. Up to that point all of the passengers, carried along in a euphoric momentum both literally and musically and seduced by the ideal form of a holiday, 'We always want to be on holiday', are happily in denial of the unobserved and unexercised powers of nature until they burst through as an electric storm. Then, just like the refugee, who is paralysed not by the storm but by his lack of identity papers and the powers they hold over mobility, they are all subject to the same reality: stuck in an airport terminal!

References

Adams, J. (1991). *The Death of Klinghoffer*. Goodman, A. [opera].

Botterill, D. (2012). Narratives of Tourism in Twentieth-century Opera, in J. Tivers and T. Rakić (eds.), *Narratives of Travel and Tourism* (pp. 77–93). Farnham: Ashgate.

Botterill, D., Pointing, S., Hayes-Jonkers, C., Jones, T., Rodriguez, C., and Clough, A. (forthcoming). Violence, Tourism, Security and Critical Realism. *Annals of Tourism Research*.

Britten, B. (1973). *Death in Venice*. Piper, M. (adaptation of Mann 1912). [opera]

Christiansen, R. (2012, 25 February). The Death of Klinghoffer: Terrorism on the High Seas. *The Telegraph*. Retrieved 15 Jan 2013 from <http://www.telegraph.co.uk/culture/music/opera/9101783/The-Death-of-Klinghoffer-terrorism-on-the-high-seas.html>.

Crawshaw, C. and Urry, J. (1997). Tourism and the Photographic Eye, in C. Rojek and J. Urry (eds), *Touring Cultures: Transformations of Travel and Theory* (pp. 176–195). London: Routledge.

Dove, J. (1998). *Flight*. de Angelis, A. [opera].

Fink, R. (2005). Klinghoffer in Brooklyn Heights. *Cambridge Opera Review*, 17(2), 173–213.

Francesconi, S. (2011). Multimodally Expressed Humour Shaping Scottishness in Tourist Postcards. *Journal of Tourism & Cultural Change*, 9(1), 1–17.

Platenkamp, V. and Botterill, D. (in press – 2013). Critical Realism, Rationality and Tourism Knowledge. *Annals of Tourism Research*, doi: <http://dx.doi.org/10.1016/j.annals.2012.12.006>.

Robinson, M. (2005). The Trans-textured Tourist: Literature as Knowledge in the Making of Tourists. *Tourism Recreation Research*, 30(1), 73–81.

The Daily Telegraph (2013, 22 January). Woman Lost Life Savings to Turkish Waiter. *The Daily Telegraph*, p. 9.

Tribe, J. (2008). The Art of Tourism. *Annals of Tourism Research*, 35(4), 924–944.

Woolcock, P. (Director) (2003). The Death of Klinghoffer. Blast! Films & Channel 4 Television Corporation (Producer). UK. [film].

PART II
Imagining, Experiencing and Remembering Places through Artwork

Chapter 6
Expeditions in Persia: Travel Journals and Artwork as Historical Documents

Mahsa Rouzrokh and Heather Mair

> The real voyage of discovery consists not in seeking new landscapes, but in having new eye.
>
> Marcel Proust, *Remembrance of Things Past*

Introduction

The relationship between art and travel is rooted in the history of civilization and the development of human societies. Long before the internet and international media organizations, the travel journals and pieces of art travellers created and transported across the globe were important ways of informing citizens about worlds other than their own. Art, ranging from written works to scenic paintings, is one such great resource that contributes to the richness of human history. Art and words are both valuable in recording and expressing the experiences of travel, however, despite the advantages of an artwork, often it is not as readily accessible compared to print as a way of conveying travel experiences.

In this chapter we focus on the importance of art in re-creating an image of the destination that is grounded in the past. Our discussion starts with a brief overview of the role of art in enriching the written documents and travel journals of the voyagers in previous centuries. Next we shed light on the important role that drawing and painting plays in the creation of one of the most valuable documents about the Persian gardens created by Donald N. Wilber. This example is important because it demonstrates the role art plays in the creation of such detailed documents about historical sites. Last, we emphasize the value of both travel journals and the accompanying artwork created by travellers in the old world as historical documents.

Sifting through the massive publications of travel accounts based on the journeys of visitors to Iran between the seventeenth and twentieth centuries requires a good deal of time and selectivity. Nonetheless, we narrowed down our focus to the works of a few travellers who precisely described Persian gardens and arrived at the famous work of Donald Wilber. Wilber's work on Persian gardens and garden pavilions published first in 1962 has valuable descriptions of Persian gardens and as such provides an excellent example for analysis. In our overview of his detailed report, we demonstrate how Donald Wilber benefited from preceding travellers' journals and drawings to create one of the most detailed documents about Persian gardens.

Iran, historically called Persia,[1] was a country of interest for foreign travellers from the seventeenth to twentieth centuries due to its rich history and spectacular art and architecture. The travellers to old Iran left behind valuable documents describing the historical sites and ruins, gardens, cultures and traditions of old Persian cities. Sketches, lithographs and drawings created by nineteenth-century travellers and commissioners, such as Pascal Coste (1787–1879), Eugene Flandin (1809–1889) and Xaviar Hommaire de Hell (1812–1848), portray the details of fountains and pools, tents and kiosks, pavilions and palaces, and of the vegetation and exuberant nature of Persian cities and gardens. In their detailed drawings, these travellers demonstrate their subjects as they experienced them at a particular moment and place in time. Moreover, their observations of their immediate surroundings are embodied in, and accurately captured by, most of the lithographs. Furthermore, most of such artwork not only focuses solely on the subject, such as gardens or pavilions, but also encompasses the larger composition of the human landscape. Foliage, flowers, water, palaces and vistas are described in great detail, as are the minutiae of the costumes, traditions and activities of local residents, which make such drawings and travel journals exceptionally valuable and incomparable as historical documents (Duncan and Gregory, 1999).

Travel, Art and Narrations: The Re-creation of the Observations

Travel is identified as a form of 'cultural encounter' (Saunders, 2012); the experience of observing cultural differences and embodying the unseen. Voyagers in the old world recorded their observations and memories in their travel journals, which accompanied them everywhere. Before the invention of photography in the nineteenth century and its popularity in the twentieth century travellers were mostly accompanied by draftsmen and artists in their voyages so that they could enrich their written descriptions with accompanying descriptive prints (Duncan and Gregory, 1999; Stafford, 1984). The origins of the travelling artist date back to the fifteenth century, 'when sketches executed on sea journeys helped supplement nautical records' (Jacobs, 1995, p. 12). A number of historians such as Williams and Hobsbawm have proposed that some travelling artists travelled to produce images and symbols in support of the national identities of both the destination and their home (Belnager, 1999). In their drawings and paintings, the travelling artists restored the richness of the scenes they were witnessing as it was experienced in the moment, re-creating the reality, and recording the history through these images. Denny (1986) and Sclater (2003) both identify the importance of art as a tool that captures representations of human life and history in a way that cannot be revealed merely through words. Art can express infinite aspects of a human experience that is

1 Please note that the terms Persia, Persians and Iran are used interchangeably throughout the text. Iran became the official, universal title for the country in 1935 (Daniel, 2012), however, both names are used interchangeably worldwide.

sometimes obscured in the language of written travel accounts (Craib, 2000; Sclater, 2003). Accordingly, scholars emphasize the important role art plays in filling the gaps between the information gathered through written history and that which can be learned from the drawings and paintings (Becker, 1991; Bochner and Ellis, 2003; Field, 1995; Stafford, 1984). The drawings and paintings created globally by voyagers in the eighteenth and nineteenth centuries also play a powerful role in the cultural identification of different countries (Belanger, 1999). The representation of a place or a site occurs in 'line and in paint, in gesture and in photograph, in image and in word' (Casey, 2002, p. xiv). As different mediums used to describe a place, drawings have played a powerful role in demonstrating both the exact details and the inherent beauty and sublimity of old sites. Even though the role of written works in describing a destination is significant, written language is not always transparent enough on its own (Field, 1995) and at times requires art to communicate with the reader. Thus, art plays an important role in helping the written word convey the experience in a more meaningful way.

Adams (1983) states that many of the travel journals published in the eighteenth and nineteenth centuries were actually revised versions of the originals that had been brought back. Some of these published travel novels are a mixture of fiction and the travellers' observations of remote sites of the world (Adams, 1983; Lamb, 2009; Stafford, 1984). In addition, Lamb (2009) has investigated the connection between the eighteenth-century novel and tourism and points out that the conjunction of travel and literature enabled writers to add imagination for flavour. Moreover, scholars have criticized the authenticity of some of the travel journals due to the embellishments later added by editors and publishers with the purpose of increasing popular reception (Stafford, 1984; Saunders, 2012). The tension existing between the first hand experience of the traveller and that of commercialized publications is especially noticeable in some of the travel narrations of the past centuries (Said, 1978; Said, 1983; Stafford, 1984; Saunders, 2012). Despite the fact that many travel writings may have a taste of fiction, the authors generally build their statements on the realities of different societies and human experiences in the time of their creation (Stevenson, 1915). In the case of drawings created by travelling artists, the works of art are mostly recording their surroundings as they were observed, documenting everything that attracted their eyes in the background. These detailed sketches, drawings and watercolour paintings entail objects, landscapes, scenes, people and even pottery shreds in a prescribed and technical way. Moreover, these travelling artists aimed at 'spread[ing] graphically before the reader and viewer a great range of exact information about the world' (Stafford, 1984, p. 59). In this regard, most of the drawings created by travelling artists can be as precise and detailed as the travel journals. The key commonality among the artwork created by the travelling artists of the previous centuries is that they portray the natural features as well as human built components of the painted scene in detail. Such detailed works of art act as historical documents, representing a detailed sense of how different sites around the world were at the time of their creation (Ambroziak, 2005). The most detailed of such works, possessing strong narrative quality, can be found in the drawings and paintings created by Coste, Flandin and Hommaire of the Persian gardens of Iran in the nineteenth century. In addition to vegetation, bridges, streams, pools and the mansions in the gardens, their paintings also depict the human landscapes, customs and traditions of the local people living in the place at the time of their travel.

These detailed travel journals and artwork created in past centuries played a role in motivating future travellers to embark on journeys to remote countries. For example, a number of devoted voyagers in the nineteenth and twentieth centuries went on expeditions to Persia and returned home with their own narratives and artwork of their explorations, building on the observations of the preceding travellers to Persia. Donald Wilber (1907–1997) was one such traveller, who developed a significant interest in traveling to Iran in the mid-twentieth century after he was exposed to the beauty of the art and architecture of Persia through the drawings and narratives of travellers such as Flandin, Coste and Hommaire. In Donald Wilber's (1962) *Persian Gardens and Garden Pavilions*, he urges the need for employing different sources of artistic representations in order to build one of the most detailed existing documents on the Persian gardens.

Wilber's Observations on Persian Gardens and Garden Pavilions

Iran is incredibly rich in cultural and heritage resources having 15 of UNESCO's world heritage sites (UNESCO, 2012). In the list of world heritage sites stands the Persian gardens, consisting of a collection of nine gardens, 'selected from various regions of Iran, which tangibly represent the diverse forms that this type of designed garden has assumed over the centuries and in different climatic conditions' (UNESCO, 2012, Par2). The unique landscape settings and architecture of these gardens amused travellers from previous centuries. Donald Newton Wilber (1907–1997), a researcher, writer and archaeologist who travelled to Iran several times in the mid-twentieth century, transformed his observations into several famous books and articles, such as Language and Society (1967), Persian Gardens and Garden Pavilions (1962) and Contemporary Iran (1963). In his book and travel journal detailing observations of Persian gardens and garden pavilions in Iran, Wilber acknowledges the need he felt to document the details of these gardens as he did not find any other source of such documentation. He describes Persian gardens spread all over the country as among the first places where he sought shelter along the long dusty road from Baghdad to Tehran on his first trip to Iran. Wilber realized that these precious gardens are either demolished or facing the threat of destruction and soon will disappear one by one. Wilber responded to this threat by creating accurate plans and drawings of the details of these gardens, a number of which did not exist any more and many were soon to disappear. His aim of creating such drawings, as he notes, was to help and encourage future generations to preserve this heritage through fostering a better and detailed understanding of these treasured gardens. In addition, in the introduction of *Persian Gardens and Garden Pavilions*, Wilber expresses his hope that such detailed documents on

Persian gardens would later be employed in the restoration of disappeared parts of these gardens. Wilber's personal response to the need to visually document these gardens was to 'make measured drawings' of the finest gardens that existed in various parts of Iran (Wilber, 1962, p. 15).

Importantly, Wilber used four sources available at that time to validate and enrich his writings and detailed drawings. The first was his own field observations as well as photographs taken by himself or his fellow travellers to Iran, second he relies on the lithographs, drawings and paintings made in the eighteenth and nineteenth centuries, third in his investigations on Persian gardens, Wilber used the information provided in Persian miniature paintings dating back to the fifteenth century; another source Wilber mentions using indirectly are the poems composed in previous centuries, which delicately described the details of some existing or demolished gardens. A number of Persian poets travelled to different parts of Persia and recorded their experiences and observations in the form of poems praising gardens and the beauty of the nature within them (Wilber, 1962). In this regard, Wilber highlights the benefit of using the poems of Sa'adi, one of the most famous poets of Iran from the medieval period, to interpret the details of the garden that were well-known in the thirteenth century and some of its ruins remains standing today:

> It is a spot beyond imagination
> Delightful to the heart, where roses bloom,
> And sparkling fountains murmur- where the earth
> Is rich with many colored flowers; and musk
> Floats on the gentle breezes, hyacinths
> And lilies add their perfume- golden fruits
> Weight down the branches of the lofty trees ... (Sa'adi in Wilber, 1962, p. 42)

Wilber also interprets the pictures of Persian gardens painted in Persian miniatures, as they were visual documents he could use to better understand the details of the layout, landscape settings and human encounters taking place in those gardens in the past. With these drawings, Persian artists created a significant impression of the atmosphere of the gardens (Wilber, 1962). In addition to the detailed drawings of the elements and plants of gardens along with traditions and costumes, events and incidents in those gardens, poems added to the margins of most of these paintings enriching the understanding of the essences of Persian gardens by providing details on the stories of those paintings. These stories provided valuable information for Wilber, as well as other investigators, about the daily life in Persian gardens (see Figure 6.1). For instance, Wilber interprets his learning from one of the Persian miniatures as:

> A painting of the early sixteenth century ... portrays a number of fairies dancing, singing, playing, and eating in their own gardens ... [another] one is that in which all or part of an elegant pavilion or palace is shown in relation to its garden. (Wilber, 1962, p. 52)

Figure 6.1 **A miniature from mid-sixteenth century**

Source: Wilber, 1962, p. 50.

In most of his observations and for the purpose of enriching his understandings, Wilber employs all sources to re-create the image of the Persian gardens. Interestingly, in Wilber's work to recreate the images of Persian gardens, both existing and vanished gardens can be found. The narrations and drawings of the vanished gardens created by previous travellers are extremely detailed and this enabled Wilber to restore the layout of some of them. For instance, Wilber describes the details of the gardens designed and constructed in Timurid dynasty in fourteenth and fifteenth centuries using the information provided by Clavijo, 'a Spaniard who went to Samarqand in 1404, describe[ing] royal audiences and festivities in the gardens in a fascinating detail' (1962, p. 54), as well as the writings of the famous fifteenth-century Persian historian, Sharaf al-din Ali Yazdi. Through the use of such detailed descriptions as well as detailed paintings of miniatures Wilber was able to reconstruct a plan of one of the vanished gardens at Samarqand (see Figure 6.2).

In the chapter on Isfahan gardens, the capital of Iran in the eighteenth and nineteenth centuries, Wilber highlights a few disappeared sections of Ali-Qapu, one of the most famous heritage sites of Isfahan. Through his studies in engravings

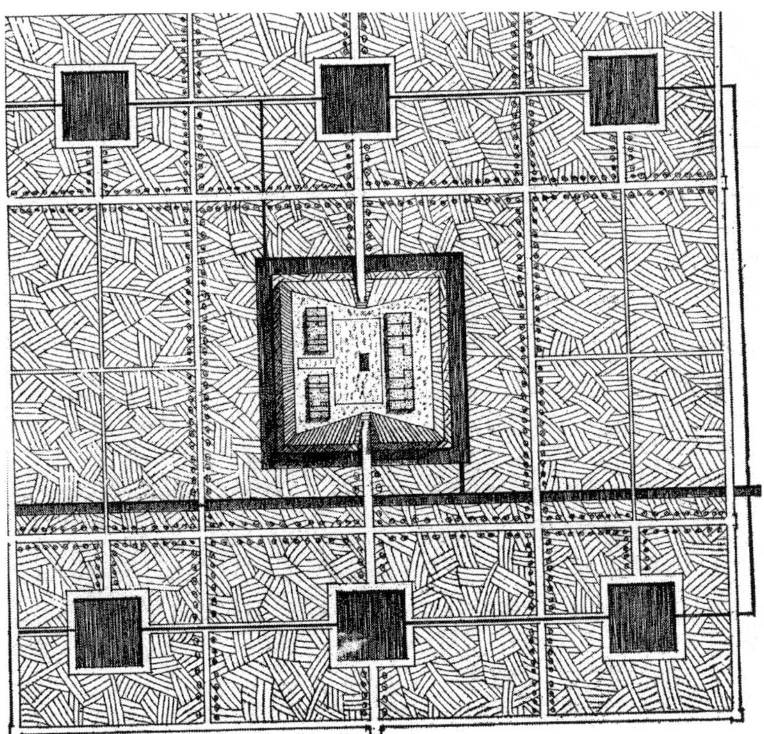

Figure 6.2 Reconstructed plan of a garden at Samarqand

Source: Wilber, 1962, p. 61.

and descriptions left from previous travellers to Isfahan, he was able to identify those displaced or vanished parts of that palace-garden. Wilber identifies a building located on the left of the palace as Talar-i-Tavileh or Hall of the Stable. He describes that the 'structure vanished in ruins more than a century ago and it is fortunate that an engraving made before 1700 and illustrating a splendid night reception has survived' (1962, p. 96). He also talks about another structure called 'Sar Pushideh' or 'Covered porch' that was erected in the opening years of the nineteenth century and did not last the century. In describing the details of this disappeared structure of the palace, Wilber builds his descriptions not only on his own observations but also on an illustration made by Coste: 'A typical garden pavilion was turned inside out; it was probably intended for occupancy during the colder months … the drawing also shows features of design and decoration which suggest that the pavilion was actually erected in the eighteenth century and merely repaired at a later time' (1962, p. 96, see Figure 6.3). Ayeneh Khane is also another vanished structure in the city of Isfahan that was once recorded with the most detail by Coste in the nineteenth century.

Figure 6.3 Engraving of the now vanished Sar Pushideh (Coste)
Source: Wilber, 1962, p. 94.

Wilber also employed information provided in travel journals. In his search for understanding the details of flowers planted and employed in the design of Persian gardens, he uses Sir John Chardin's *Travels in Persia*. Chardin, who travelled to Persia between 1673 and 1677, describes his observations on the roses planted in Persian gardens in excellent detail:

> The rose is found in five colors: white, yellow, red, Spanish rose, and poppy red. Also there are two faced roses which are red on one side and yellow on the other. Certain rose bushes bear yellow, yellow-white, and yellow-red roses on the same plant. (Chardin, 1711 in Wilber, 1962, p. 23)

In describing the gardens of Tehran and the ones constructed along Caspian Sea, Wilber benefits enormously from the drawings and sketches created by Flandin and Hommaire. From the descriptions provided by Chardin and Sir Thomas Herbert, who travelled to Persia between 1627 and 1629 and visited the northern city of Iran, Mazandaran, along with the sketches and drawings created by Hommaire in the eighteenth century, Wilber was able to reconstruct the plan of the vanished parts of Ashraf garden in northern parts of Iran (see Figures 6.4 and 6.5). Herbert describes the details of one of the vanished palaces through his memory of visiting it and in his published travel journal of his travels to some parts of Asia and Africa: '... far from the terraces were viewed the Caspian sea ... The ground chambers were large, quadrangular, arch, and richly gilded above and on their sides'(Herbert, 1638 in Wilber, 1962, p. 135).

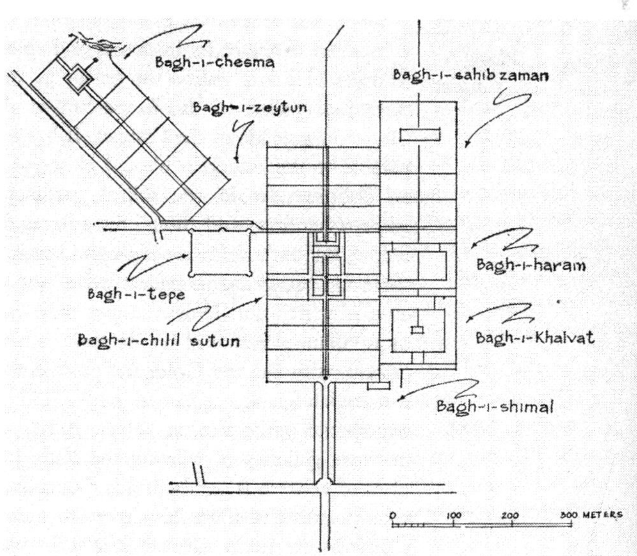

Figure 6.4 Restored plan of Ashraf gardens
Source: Wilber, 1962, p. 128.

Figure 6.5 Engraving of vanished garden palace at Ashraf (Hommaire) 1
Source: Wilber, 1962, p. 124.

In studying the palace and gardens of Gulestan in Tehran, Wilber overlaps the descriptions provided by French traveller Pierre Amédée Emilien Probe Jaubert (1779–1847) on this palace and garden in 1806 and Robert Ker Porter's (1777–1842) trip in 1821 as well as the paintings created by Flandin in 1839–41. From the paintings Wilber was not only able to describe the element of the gardens but also the paintings created by Hommaire, which portrayed the typical Tehran house of the nineteenth century; complete with courtyard, pool and garden as well as the daily life of its residents (see Figure 6.6).

Conclusion

Recent evidence confirms that some of the published travel narratives heavily rely on descriptions provided by earlier travellers (Jacobs, 1995). The detailed descriptions provided in travel narrations add significant depth and context to the understanding of

Figure 6.6 Engraving of vanished garden palace at Ashraf (Hommaire) 2
Source: Wilber, 1962, p. 144.

future travellers of different places in the world. Although some scholars debate the originality of some of the published travel journals (Adams, 1983; Said, 1978; Said, 1983; Stafford, 1984; Saunders, 2012), they agree on the value of the information provided by such narrations. In addition, the drawings created by travelling artists accompanying most of these voyagers' journals illustrate solely the observations of travelling artists, having the potential to be a travel journal itself. Although the viewer of these pictures must bear in mind that many of the finished paintings were also developed after the artists' return to their home countries (Jacobs, 1995), the nature of such paintings demands the artists in the early nineteenth century to re-create representations of the places and landscapes as they were at the time (Gombrich, 1950). Such works of art, powerful in representing the details of the sites, helped travellers and researchers such as Wilber to create a comprehensive historical document of different heritage sites of the world.

In this era of instantaneous communication and accessible travel, new meanings are introduced to the world of travel and art (Bonami, Widholm, and Van

Eck, 2005). Even though the means of recording the observations and storing the experiences give the travellers many opportunities to bring back a vast selection of their observations and experiences, the delicate details embodied in the travel journals and sketches created by tourists and explorers still play a significant role in enhancing our understanding of travel sites (Urry, 2002). Despite the fact that the invention of cameras enables travellers to capture the moments of their travel with inclusive detail, scholars such as Jacobs (1995) believe that the beauty of the scene can only be captured through the subjective eyes of the painters. Travellers of previous centuries were more than just voyagers. They can be considered scientific explorers who were both writers and artists exploring parts of the world that were inaccessible to the majority of the population. Through their explorations, these exceptional travellers created documents that recorded evidences of parts of the world in the past centuries, the landscapes, marvellous scenes and traditions that do not exist any more. Although many of the authors of published travel journals between the eighteenth and twentieth centuries may have struggled to re-present the reality of their observations without the intervention of the publishing world (Stafford, 1984), the illustrations they have provided in their works leave less room for imagination by re-creating the picture of the observed scenery.

As can be understood from the work of Donald Wilber, journals, travel books and diaries accompanied with sketches, drawings and paintings created by tourists in the previous centuries are more than just travel memoirs and art. They are historical documents illustrating the natural, historical and cultural settings of the old world. These illustrative materials, works of art so precise, delicate, detailed, and faithful to life were brought back to the travellers' home countries and later were spread all around the world. Moreover, this artwork was marvellously successful in capturing the atmosphere of living in a specific time and place on the other side of the world. Not only was the location the matter of record, but the social encounters and interactions were also captured and recorded (Clifford, 1997), as travellers of the old world looked not just at their surroundings, but saw, absorbed and experienced every detail of living in those parts of the world.

References

Adams, P. (1983). *Travel Literature and the Evolution of the Novel*. Lexington: University of Kentucky Press.

Ambroziak, B. M. (2005). *Micheal Graves: Images of a Grand Tour*. New York: Princeton Architectural Press.

Becker, A. L. (1991). A Short Essay on Languaging, in F. Steir (ed.), *Research and Reflexivity* (pp. 226–234). Thousand Oaks: Sage.

Belanger, P. J. (1999). *Inventing Acadia: Artists and Tourists at Mount Desert*. Rhode Island: Farnsworth.

Bonami, F., Widholm, J. R., and Van Eck, T. (2005). *Universal Experience: Art, Life, and the Tourist's Eye*. Chicago: Museum of Contemporary Art.

Bochner, A. P. and Ellis, C. (2003). An Introduction to the Arts and Narrative Research: Art as Inquiry. *Qualitative Inquiry*, 9(4), 506–514.

Casey, E. S. (2002). *Representing Place: Landscape Painting and Maps*. Minneapolis: The University of Minnesota Press.

Clifford, J. (1997). *Routes: Travel and Translation in the late Twentieth Century*. Cambridge, MA: Harvard University Press.

Craib, I. (2000). Narratives as Bad Faith, in M. Andrews, S. Day Sclater, C. Squire, and A. Treacher (eds), *Lines of Narrative* (pp. 67–74). London: Routledge.

Daniel, E. L. (2012). *The Greenwood Histories of the Modern Nations: The History of Iran*. Santa Barbara: Greenwood.

Denny, J. M. (1986). Art Therapy in the Counseling of a University Student. *Bulletin of Art Therapy*, 7(2), 94–99.

Duncan, J. and Gregory, D. (1999). Introduction, in J. Duncan and D. Gregory (eds), *Writes of Passage: Reading Travel Writing* (pp. 1–13). London: Routledge.

Field, R. S. (ed.) (1995). *Mel Bochner: Thought Made Visible 1966–1973*. New Haven: Yale University Art Gallery.

Gombrich, E. H. (1950). *The Story of Art*. London: Phaidon Press Ltd.

Jacobs, M. (1995). *The Painted Voyage: Art, Travel and Exploration 1564–1875*. London: British Museum Press.

Lamb, S. (2009). *Bringing Travel Home to England: Tourism, Gender, and Imaginative Literature in the Eighteenth Century*. Cranbury: University of Delaware Press.

Said, E. W. (1978). *Orientalism*. London: Penguin.

Said, E. W. (1983). *The World, the Text, and the Critic*. London: Harvard University Press.

Saunders, A. (2012). Posting Overseas: The Travelling Tales of Anthony Trollope, in J. Tivers and T. Rakić (eds), *Narratives of Travel and Tourism* (pp. 109–120). Abingdon: Ashgate.

Sclater, D. S. (2003). The Arts and Narrative Research – Art as Inquiry: An Epilogue. *Qualitative Inquiry*, 9(4), 621–624.

Stafford, B. M. (1984). *Voyage into Substance: Art, Science, Nature, and the Illustrated Travel Accounts, 1760–1840*. Cambridge: MIT Press.

Stevenson, R. L. (1915). *Essays of Travel and in the Art of Writing*. New York: Charles Scribner's Sons.

UNESCO (2012). *The Persian Gardens*. Retrieved on 05 September 2012, from<http://whc.unesco.org/en/list/1372>.

Urry, J. (2002). *The Tourist Gaze*. (2nd ed.). London, UK: Sage.

Wilber, D. N. (1962). *Persian Gardens and Garden Pavilions*. Tokyo: Charles E. Tuttle.

Wilber, D. N. (1967). Language and Society: The case of Iran. Cross-Cultural Research, 2(1), 22–30.

Wilber, D. N. (1963). Contemporary Iran. New York: Praeger.

Chapter 7
Carving the Past: Art, Tourism, and Heritage in Chincoteague, Virginia

Kristin M. Sullivan

Introduction

In cultural anthropology the art market has long been understood as related to social structure, with different modes of production related to different relationships between producers and outsiders such as tourists. Less attention has been paid, however, to the ways in which art is linked to a tourist (host) community's heritage. My research examines art's connection to heritage in a touristic setting. Hunting decoy and decorative wildfowl carvers in the resort town of Chincoteague, Virginia (USA) are one group of utilitarian craftsmen-turned-artists whose work is intimately tied with their community's past, and understandings of their evolving community today. Wildfowl carvers, veterans of a rural lifestyle involving waterfowl hunting and trapping, have continued and passed on a vernacular art form that reflects local values associated with work, the environment, and other forms of cultural and natural heritage. They have done this despite the gradual decline of access to natural resources due to increasing government presence, environmental pressure, and tourism growth. The presence of decoys within particular tourist markets, the carvers' selective participation in island events and organisations, and the ways in which carving is learned and passed on reveal much about the ways in which art is used to express and continue heritage, and negotiate identity.

This chapter provides a review of relevant literature, and then explores some of the ways in which hunting decoy and decorative wildfowl carving in Chincoteague has evolved from its previous utility into an art form marketed for tourist consumption. I will show that those involved in the sale of carvings in the region have managed their heritage by managing the production and performance of their art, a notion that lies outside or on the periphery of early theoretical literature within the anthropology of art and tourism, which tends to downplay artist and local community agency and leave heritage almost entirely out of the discourse.

Setting the Scene

Three life-size wooden carvings of ducks, called decoys, sit in a storefront window in downtown Chincoteague Island. These decoys are placed atop a mirror, apparently representing water. Behind them is a framed painting of the Assateague lighthouse, a landmark present since the mid-1800s only a few miles away, on what is now a federally protected wildlife refuge. Surrounding these objects are baskets of ivy, and carved waterfowl placed on driftwood. Many tourists see this display as straight out of the refuge, beautiful birds as if in their natural environment, invoking a sense of serenity or a connection with a public natural heritage. Over the past couple years I have come to know this scene as something a little different.

While the scene is certainly serene, it is born of hunting traditions from which the decoys come, and speaks to a local heritage of decoy carving as necessary for attracting wildfowl for food, and survival. The shop owner is likely aware of this, and yet has positioned the hunting decoys in a simulacrum of nature quite on purpose for his tourist audience, perhaps playing to their expectations. Decoy and decorative wildfowl carvers, shopkeepers, and others involved with this art form in the Chincoteague tourism industry have managed their local heritage in such a way that the once primarily utilitarian craft of decoy carving has become a means to position oneself in the potentially lucrative tourist art market and, at the same time, convey an important part of their heritage.

Situating Chincoteague

Chincoteague is a small barrier island off the coast of Virginia that has become known for its seafood and waterfowling, as it is surrounded by productive salt marshes and brackish bays, and is positioned along the Atlantic Flyway, a bird migration route following the east coast of North America (see Figure 7.1, Chincoteague Island, with Assateague Island and the Atlantic Ocean in the distance). Industries revolving around these resources long ago took root, and many long-time local residents, 'Chincoteaguers', trace their lineage along generations of watermen, trappers, and hunters. Nevertheless, Chincoteague today finds itself less a fishing village and more a modern tourist destination with visitors attracted to the picturesque town and to nearby Assateague Island's federally protected seascape, which is connected by bridge to Chincoteague. Many long-established commercial fishing and hunting practices are becoming repurposed and replaced by tourism related activities as long-time residents find it increasingly difficult to make a living and as new residents with new goals, many of whom are retirees, slowly become the majority. Complicating this, Chincoteague's major tourism assets are today controlled by the United States Fish and Wildlife Service (USFWS) and the United States National Park Service (NPS). The natural resources of Assateague Island have helped make tourism a $24 million per year industry on Chincoteague and provide habitat for a range of wildlife species (Mariner, 2010). While the resources are thus arguably economic and environmental boons to the island, the agencies that manage them are often perceived as threats to local independence and to the continuance of traditional life-ways, as the now-protected lands include traditional hunting and fishing grounds as well as traditional places of recreation.

Figure 7.1 Chincoteague and Assateague islands
Source: Photo by Patrick J. Hendrickson; www.highcamera.com, used with permission.

Decoy carving on Chincoteague is a tradition that dates back long before the federal agencies were put in place, to a time when wooden decoys (also called gunning stools) were used to fool live ducks, geese, and shorebirds into thinking hunting grounds were safe for landing. Carvers would create rigs of decoys for hunting purposes, for their own use, or for others to buy and use personally or as a hunting guide. Hunting parties were composed of some of the region's earliest tourists. Affluent men from nearby cities such as Baltimore, MD and Washington, DC would visit to hunt for sport, using local guides and locally made decoys. Yet even then hunting decoys were given aesthetic consideration and even considered art. One bird from a rig might be saved by someone in a party, and placed on a shelf back home as a souvenir.

Into the mid-twentieth century gunning decoys were sold for no more than $1.25 US. A few noted carvers were sought out for their work, however, and reputations grew. By the 1960s many Chincoteague carvers were creating gunning-style birds for sale as decoration and taking special orders from around the country. Also by the 1960s both the Chincoteague National Wildlife Refuge (USFWS) and the Assateague Island National Seashore (NPS) were in place, along with increased state hunting restrictions. Between the reputations gained by a handful of fine carvers, the diminished access to hunting grounds, and the development of mass-produced plastic decoys for hunting purposes, carved wooden decoys shifted in purpose. What was once primarily used for hunting and occasionally art became known as an art form occasionally used for hunting. Today wildfowl carving has developed into a highly refined artistic tradition, bringing thousands of dollars per bird to top carvers, and thousands of tourists to local carving competitions, shows, and shops.

Background and Research

I first visited Chincoteague in the height of the tourist season, a hot summer day buzzing with people in shops, ice cream parlours, and on their way to the beach. I wandered into a garage-turned-decoy shop on the side of a carver's home, off of the boulevard leading to Assateague Island. After looking around for a while I found a bird off to the side that I particularly liked for its roughness, its gestural appearance, and other aesthetic and natural qualities I value. The shop owner, also the carver in residence, quickly chastised me for my choice, saying the bird was not finished; it was not yet for sale. He explained to me the qualities it still lacked, the same qualities that would make it a good decoy: the proper colouring, certain wing and eye details. He followed up on this lesson with stories of his father carving, and a brief history of hunter-carvers on the island. I was intrigued by this turn of events. Here was this beautiful object, in my estimation, that he understood as needing to be refined in order to meet his standards of appropriate art before he would sell it. He was able to sell me another bird that met both our aesthetic criteria; however, in refusing me the first bird for the reasons he gave, he placed himself in a tradition of carving related to his community's heritage, a heritage of working the land and water, of using natural resources for food, and of taking pride in the objects that were perhaps meant only to be shot over in the process. He also managed how that heritage was distributed.

In the two years since this first encounter I have had the opportunity to return to Chincoteague on a number of occasions, both for field visits during a study of regional heritage for the NPS, and for my own on-going dissertation fieldwork in anthropology. During these visits I have relied heavily on qualitative methods such as participant observation with carvers, collectors, tourists, and museum staff; interviews with carvers, collectors, shop owners, and tourists; and text analysis of oral histories and previously collected interviews with carvers. The following analysis is based on this preliminary research and remains on-going.

Art and Tourism in Anthropology

In cultural anthropology, tourist art has often been understood as a means to understand social structures. Less attention has been paid to art's relationship to the ways in which locals and outsiders understand a place's and people's past, and artists seem to possess little agency. Many writers in the 1970s and 80s suggest that tourist art is part of social integration and differentiation (Graburn, 1976), where locals may produce art separately for tourists and insider populations (external and internal audiences); but tourist art is yet a reflection of tourists' desires, or artists' responses to forces in the tourist market (Abramson, 1976; Low, 1976; Deitch, 1989). Art is thus a potentially hegemonic marker by which we might understand seemingly powerless people (Graburn, 1976, 1977) in relation to the outside world and its *tourist gaze* (Urry, 2002), rather than as a means to understand how artists craft their own links to their consumers and their past.

Increasingly, recent ethnographic accounts concerned with art and tourism do address the roles artists and their surrounding communities play in the production and sale of their work. No longer mere descriptions of stylistic and social changes resulting from so-called tourist impact, these studies address concerns such as identity formation and representation (Adams, 2006); shifting gender roles (Tice, 1995); human agency, work, and creativity (Colloredo-Mansfeld, 1999; Duggan, 1997); questions of authenticity (Littrell, Anderson, and Brown 1993); and the effects of anthropology on the study of tourist art (Castañeda, 2005). The insistence on framing art and tourism within a structuralist model persists, but generally this frame is used to consider greater issues of power negotiation and the ways in which artists may move within and move out of roles as result of economic success, fame, and other forms of cultural capital.

Notably, Adams (2006) writes of how art is something that can be activated to display self-attributed values. She suggests that art has '"an affecting presence" imbued with emotional force', and a space with art provides 'a particularly apt arena for negotiating, reaffirming, and at times challenging … social identities' (2006, p. 27). Part of what Adams and others exploring identity seem to be talking about are the ways in which art is used to present and represent identity found through heritage, and yet the concept of heritage is rarely mentioned in the anthropological literature on art and tourism. On the other hand, tradition, often associated with heritage, is considered at length (Moreno and Littrell, 2001).

Tradition is often seen as the opposite of modernity; and due to the same modern factors that motivate the *tourist gaze* or the desire to see the authentic *back stage* of a tourist setting (c.f. MacCannell, 1999), there exists a modernist desire to preserve tradition, the remnants of some authentic past. Nostalgia drives the present to idealise this past (be it fictitious or otherwise); and for scholars studying art in tourism, the drive to study or preserve tradition leads to near obsession over stylistic details, materials used, explaining the ways in which these changes are no longer traditional and the ways in which tourism has had an impact on tradition. Less emphasis is placed on the meaning traditional elements hold from an emic perspective, or why a local (host) population might change tradition over time. I believe it is important to consider not just the traditional elements of art, but the art's, the artist's and host community's, heritage in studying tourist art, in order to understand how their heritage is managed through the tourist and art industries.

Art, Tourism and Heritage

Heritage involves something akin to claiming identity through the past. Not quite history, as it has a mutable life in the present, heritage is a varyingly tenuous link to history that is intended to continue into the future, giving meaning to those who claim it. Heritage claims can legitimate a place, person, or thing, providing meaning and value (c.f. Kirshenblatt-Gimblett, 1998). I would argue that one could also begin to understand heritage and the meanings and values associated with it through art in touristic settings.

Tourist art takes on a range of meaning for locals and tourists that can be expressed in terms of what may be understood as public and private heritage, where the former involves the attempt to preserve an idea of what the past might have meant, and the latter looks at the past through its continuity in a living community (c.f., Chambers, 2006). Private heritage may further be broken down into an assemblage of personal heritages, the veins of continuity running through individuals within a community body of tradition. On one hand, if we consider the private and personal heritages of the arts we may ask several questions: In what traditions has an artist been taught stylistically, what skill sets have been learned, what gender or other identity roles are assumed, etcetera? What is the significance of the pedagogical design, of teaching and learning the art, but also what is the art intended to tell the viewer? How does the art relate to the artist's worldview and understanding of the artist's past, and how is the past represented? On the other hand, if we consider the public heritage of the arts we may ask similar questions with regard to artistic communities, representations of worldviews, and so forth; but we must consider additional questions: Who decides its meaning, and for what purposes is it created and displayed? What does its display tell us about social and political positioning? Museums and other public heritage venues run the risk of creating 'a kind of "orientalism", a fascination with the "otherness" of other cultures, times, and places, in exhibitions of all kinds that deal in *absent* people' (Dicks, 2003, p. 145), absent not necessarily due to physical location or time, but absent from majority or popular discourse as well. Public heritage displays similarly run the risk of creating the impression of a unified people or ethnicity. People may be lumped in to a certain heritage (for example, native peoples, farmers, craftsmen, rural Americans) and presented either as of one people, or as an unproblematic cultural mosaic (Chibnik, 2006). Artists expressing their private or personal heritage may find their art repurposed in public heritage contexts in such a way that emphasises certain broad themes important to a large public (or region, nation, cause) while deemphasising connection to the maker's personal past (which may even be in opposition to the public heritage, for example a hunter-carver's association with killing, the thing a tourist buys a piece of art to remember). While this may easily be viewed in a negative light, where public heritage institutions (for example, museums) co-opt or misappropriate heritage, an artist may also take advantage of the divide in forms of heritage as well, creating and marketing objects according to an audience. To be sure, the shopkeeper's display described at the beginning does just this; it takes private heritage objects and situates them in such a way as to appeal to a public heritage-seeking audience. The hunting decoys have become part of the story of one version of wildlife protection; they are a simulacrum of nature in situ.

Art, Tourism and Heritage in Chincoteague

Chincoteague carvers are acutely aware of the different audiences for which they carve, as well as the different ways other carvers use Chincoteague's heritage. As one carver informed me tongue-in-cheek, 'there are carvers, and there are trinket-makers'. When asked what makes a carver a carver, he replied that they have to have lived a certain lifestyle, have hunted, 'been out there'. Trinket makers, on the other hand, have capitalised on the limited successes of decoy carvers, creating decorative birds called decoratives, fit for public consumption, in line with ideas about the public heritage of Chincoteague. What this particular carver refers to in discussing a sort of authentic carving may be what Chiarappa calls 'vernacular craft':

> Vernacular craft is defined as artefact production that embodies the functional, historic and symbolic relevance of a community or region's occupations, social relations and environmental interaction. Furthermore, the producers of these crafts are community members who have, from their own life experience, a deep knowledge of the history and cultural logic of these issues. (1997, p. 339)

Decoy carvers aren't artists who simply carve and paint what they see. Carvers draw on their private and personal heritages, heritages that involve intimate contact with their surrounding land and waters. They draw on a body of environmental knowledge and set of values that relate to uses of the environment, and they have an understanding of the ways in which their art has been used practically and politically in the past (that is, its utilitarian value and history). Importantly, carvers tend to learn to carve by observing other carvers. I have yet to meet a Chincoteague carver who underwent formal training for the purposes of carving decoys, including formal apprenticeship. Instead, knowledge is passed on through informal transfer and learned through observation. Carver Roe Terry recalls his own education in a magazine interview, with commentary from the author:

> 'When I grew up, I wanted to be just like [decoy carver and waterman Doug Jester, Jr.].' … Terry wasn't 10 years old when he first went hunting with Jester, taking his mentor's 10-gauge double barrel shotgun and a set of shorebird decoys. 'He did the shooting and I did the retrieving,' [Terry] remembers. … After school, when Jester wasn't working on the water, Terry enjoyed going to his shop to watch him work. 'He was always fixing and repairing things, always talking as he worked,' Terry says. 'I got to watch him cut out and make gunning decoys, as well as decoratives, which he usually gave away. (Trimble, 2006, p. 41)

This passing of knowledge from generation to generation in a subtle, informal way, with respect due to earlier carvers, is very much part of the continuance of heritage (see Figure 7.2, Chincoteague carver Jimmy Bowden taking a break from his work to tell a story about carving). It is also evident in the art object produced. Several modern Chincoteague carvers report having spent time with Miles Hancock, a renowned carver and entrepreneur from an earlier generation. Most of them have shown me decoys made in his style; they learned to carve like Hancock before adapting and building on Hancock's techniques to create their

Figure 7.2 Jimmy Bowden in his workshop

Source: Author's own photograph.

own style. Similarly, one of Chincoteague's most famous carvers, Delbert 'Cigar' Daisey, who himself learned to carve alongside his father, allowed a handful of young men to watch him carve, and he reports that he can see his influence clearly in their styles. He and other carving enthusiasts can pick out which birds were made with his patterns and techniques, versus the ones created by the younger carvers alone. In this way there is physical evidence of intangible inheritance. Skill sets, embodied cultural understandings of nature, and values are passed on from generation to generation, with current carvers and their work representing their heritage. Carving is a performance of embodied cultural knowledge. Even when skill lacks evidence of education, artwork may still be evidence of the past, in what the object was once used for. Chincoteague carver Bobby Umphlett tells how he began carving:

> Umphlett: The reason that I got into it: I wanted to hunt, and were poor. I wanted to make my own stuff, and I started, and oh they were crude. (laughs) But, as I got going I got a little better, and the tourists started buying them! So I'd make

> them to hunt with that winter, and the next summer I'd end up selling them and had to make all new ones. But, the money was good, and that's how I got doin'. I started hunting. …
>
> Kristin Sullivan: Who'd you learn to carve from when you were first making them to hunt with?
>
> Umphlett: Me. Nobody taught me. I got my hatchet out and started just trying it. … I learned basically myself. Then other boys got into it and we'd exchange. You know, something I liked that he'd done; we'd kind of talk back and forth like that. That's how we did. …
>
> Kristin Sullivan: It doesn't sound like anybody was ever really formally taught … .
>
> Umphlett: We kind of picked up. Well, if you wanted to hunt, and you wanted some decoys, that was probably the best way to do it – get you a block of wood. And some of these decoys now, they're expensive! (Interview with the author, 11 October 2012)

Today many of Umphelett's carvings exhibit fine detail and lifelike features. These birds are not decoys but decoratives, meant for display purposes only. Nevertheless they are in a line of water fowling tradition, where huntsmen were valued for their ability to put food on the table and maybe turn over a dollar or two by selling their product to a tourist. Carving was necessary for their way of life. Thus heritage continues and is present in the art object (the decoy or decorative) as it is evidence of the tradition (carving lineage and skill set, or hunting tradition) and it is evidence of the necessity that brought it to fruition.

Situating and Selling Chincoteague's Heritage

Carved ducks, geese, and other wildfowl are today found throughout the island on mailboxes, as lawn ornaments, on shop signs, and so forth. Decorative birds are sold in many gift shops, but they are largely made for export elsewhere in the world, sometimes molded of plastic, resin, or other materials uncommon in traditional carving. It is telling to see which shops feature different types of decoys and other carved waterfowl. Resin decoratives are predominantly in shops that also sell beach towels and bumper stickers, or other common tourist souvenirs found in many beachside resorts. Most local carvers' works are on display either at their home workshops, or in specialised stores that seem to be more in line with the carvers' ideas about proper tourism development.

I am told that tourists will buy small, inexpensive decoys in stores, but that it is a rare tourist who will seek out a well-made, and therefore expensive, decoy by a Chincoteague carver in a tourist store. In the early 1970s tourists visiting Chincoteague's 'Visitor Information' stands would receive 'a schematic map of the Island with a listing of the local decoy carvers and their addresses' (Berkey and Berkey, 1981, p. 87) (see Figure 7.3, a map of Chincoteague decoy carvers by Evelyn Taylor, ca. 1970). Today some of the carvers featured on that map still sell their work, but rarely in stores, and never with advertisement. Their art, still sought after, is limited in distribution, perhaps to protect its heritage value.

For example, if you want to buy one carver's specialty you find the street where he lives and follow the hand-painted signs to his backyard workshop. Another carver sells his decoys out of a small stand on his front lawn on a well-travelled street (see Figure 7.4). These two artists have responded to the desires of different audiences in different ways very much on their own terms, which is evidence of their ability to adapt to and make decisions about varying tourist art markets. The first of these carvers sells decoys out of the same shop in which he creates them. He has no set hours and relies on his reputation for sales. He regularly stocks a small number of birds for quick sale, but relies in large on special orders from tourists who have found his shop over the years by word of mouth. It should be noted, however, that the first bird of each species this carver makes is given to his wife. He places high personal heritage value on some of his pieces, and distributes them accordingly. The second carver creates hundreds of small birds for sale at low prices, which he sells on his front lawn. He creates decoys for an external audience, but retains aspects of traditional carving and painting common in his

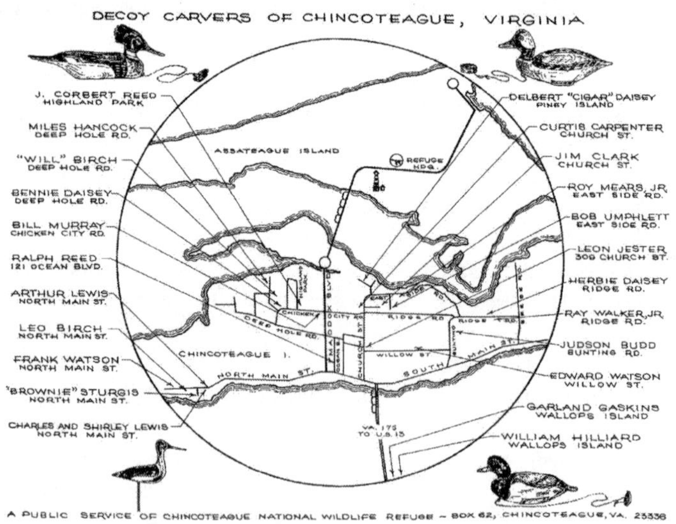

Figure 7.3 Map of the carvers

Source: Courtesy of the Museum of Chincoteague Island, reproduced with permission.

family. He creates and sells objects related to the island's private heritage in a way that appeals to the masses of tourists, creating a true tourist *borderzone* (Bruner, 2005), a leisure space where tourists have access to the 'exotic' Other, that locals, on the other hand, use as spaces of commerce away from their true *back-stage* lives.

A third Chincoteague carver is fully aware of his various audiences and capitalises on the opportunities related to each. On one visit to his workshop he pulled out three bufflehead drake (male duck) decoys for me to examine (see Figure 7.5). One is made in the style he most naturally gravitates toward making. It is well constructed, simply painted, and is able to serve as a decoy but is intended as decoration. The second started its life as the exact same decoy, but he took the time to rough up the paint, and screw in and then take back out a metal plate on the bottom and attach an eyehook in the front, to give the impression that it had been used in hunting as part of a rig. The third uses the same carving pattern as the first two decoys, but is painted in greater detail, waxed to a smooth sheen, and has metal on the bottom to balance it properly. This third decoy is for competition in a decoy carving contest. I asked the carver how people respond to the different birds, and he said that consumers at decoy shows, and tourists who find him, tend to gravitate toward the antiqued decoy. In fact, he mostly sells decoys that he's purposefully made to look old. There are a few people who will buy the first style; these tend to be local consumers. The third style isn't meant for sale; it is meant to establish the artist according to public set standards.

Figure 7.4 Herb Daisey's roadside decoy shop

Source: Author's own photo.

The second, however, sells well and shows us that tourists do tend to gravitate toward an idealised past represented and authenticated by a local person tied to that past (perhaps public heritage in the guise of private), rather than gravitating toward the art object most related to the living connections to the past (private heritage). Regardless, it is the carver who negotiates his market. He understands the various demands, and adjusts designs accordingly. None of what he does falls outside of what he wants to do; he makes his living elsewhere and does not rely on tourist art for sole income. He creates what is appropriate for him in different situations. Through the decoys and decoratives made by these individuals we can see lineages or legacies of carving tradition or heritage, but we can also see agency in adaptation to tourist markets, and active control over the marketing and distribution of the artist's product.

Figure 7.5 Bufflehead decoys

Source: Author's own photo.

Conclusions

In *Zen and the Art of Motorcycle Maintenance* Pirsig (1974) describes what he calls 'mechanic's feel', a sort of incorporated knowledge that an action is right, apparent when practicing a finely honed skill, manipulating wood with finesse, for example. I associate this with habitus, with an ability to perceive, know, and react with one's body based on years of training, or being in a given culture. I am convinced that carvers understand this implicitly. Several times I have asked carvers how to start carving birds. Nearly every time the response I receive is, 'You just take away everything that doesn't look like a bird'. There are books written about carving, but most carvers I have met thus far have learned through trial and error, observation of other carvers, and close association with the birds themselves in their natural environment. Carvers understand what the birds look like from endless hours watching them in their backyards or from hunting blinds, and they know how to carve because they have been doing it since they were young, and they learned from watching their father or neighbour, who did the same, for generations. For Chincoteague carvers there exists a very personal heritage of connection to the land and water, and perhaps especially to the community that thrives on these resources. Importantly, however, there is also a sense that this heritage can be managed through the production and sale of decoys and decorative wildfowl. Over the years in Chincoteague, carving has evolved from a primarily utilitarian (though skilful) task to an artist's arena. Tourist shops have opened that attempt to capitalise on the popularity of decoy carving, but the local carvers of gunning style birds and fine decoratives alike have decided how their style will evolve, and the ways in which their sales will take place: in locally-run shops, from their back yards, in bulk, by special order, and so on. There is no one size fits all tourist art formula for Chincoteague decoy carvers, and I think this is in large part due to the personal heritage connection these carvers have to those who taught them, the reasons for which they began carving, and the ways they feel are fit to pass on such an important and on-going part of their past to those tourists and collectors who become interested in their craft. As such, Chincoteague decoy and decorative wildfowl carvers illustrate well the ways in which heritage is strongly tied to tourist art, and the extent to which artists act as agents on behalf of that heritage. This research contributes to the understanding of the relationships between heritage and tourist art and the modes in which these are intertwined through transmission of skill and active engagement with different art markets. It also points to the need for further exploration of the ways in which host community artists are not 'powerless' or passive members of a tourism industry, but powerful actors, shaping their identity and negotiating heritage values through the production and distribution of tourist art.

References

Abramson, J. A. (1976). Style Changes in an Upper Sepik Contact Situation, in N. H. H. Graburn (ed.), *Ethnic and Tourist Arts: Cultural Expressions from the Fourth World* (pp. 249–265). Berkeley: University of California Press.

Adams, K. M. (2006). *Art as Politics: Re-crafting Identities, Tourism, and Power in Tana Toraja*. Honolulu: University of Hawai'i Press.

Berkey, B. and Berkey, V. (1981). *Chincoteague Carvers and Their Decoys*. Gettysburg: Herff Jones University.

Bruner, E. M. (2005). *Culture on Tour: Ethnographies of Travel*. Chicago: University of Chicago Press.

Castañeda, Q. E. (2005). Between Pure and Applied Research: Experimental Ethnography in a Transcultural Art World. *NAPA Bulletin*, 23, 87–118.

Chambers, E. (2006). *Heritage Matters: Heritage, Culture, History and Chesapeake Bay*. College Park: Maryland Sea Grant College.

Chiarappa, M. J. (1997). Affirmed Objects in Affirmed Places: History, Geographic Sentiment and a Region's Crafts. *Journal of Design History*, 10(4), 399–415.

Chibnik, M. (2006). Oaxacan Wood Carvings in the World of Fine Art: Aesthetic Judgments of a Tourist Craft. *Journal of Anthropological Research*, 62(4), 491–512.

Colloredo-Mansfeld, R. (1999). *The Native Leisure Class: Consumption and Cultural Creativity in the Andes*. Chicago: University of Chicago Press.

Deitch, L. I. (1989). The Impact of Tourism on the Arts and Crafts of the Indians of the Southwestern United States, in V. L. Smith (ed.), *Hosts and Guests: The Anthropology of Tourism* (2nd ed., pp. 223–236). Philadelphia: University of Pennsylvania Press.

Dicks, B. (2003). *Culture on Display: The Production of Contemporary Visitablity*. London: Open University Press.

Duggan, B. J. (1997). Tourism, Cultural Authenticity, and the Native Crafts Cooperative: The Eastern Cherokee Experience, in E. Chambers (ed.), *Tourism and Culture: An Applied Perspective* (pp. 31–57). Albany: SUNY Press.

Graburn, N. H. H. (1976). Introduction: The Arts of the Fourth World, in N. H. H. Graburn (ed.), *Ethnic and Tourist Arts: Cultural Expressions from the Fourth World* (pp.1–32). Berkeley: University of California Press.

Graburn, N. H. H. (1977). Tourism and Ethnic Arts. *Annals of Tourism Research*, Oct/Dec, 53–57.

Kirshenblatt-Gimblett, B. (1998). *Destination Culture: Tourism, Museums, and* Heritage. Berkeley: University of California Press.

Littrell, M. A., Anderson, L. F., and Brown, J. (1993). What Makes a Craft Souvenir Authentic? *Annals of Tourism Research*, 20, 197–215.

Low, S. M. (1976). Contemporary Ainu Wood and Stone Carving, in N. H. H. Graburn (ed.), *Ethnic and Tourist Arts: Cultural Expressions from the Fourth World* (pp. 211–226). Berkeley: University of California Press.

MacCannell, D. (1999). The Tourist: A New Theory of the Leisure Class. Berkeley: University of California Press.

Mariner, K. (2010). *Once Upon an Island: The History of Chincoteague* (3rd ed.). Onancock: Miona.

Moreno, J. and Littrell, M. A. (2001). Negotiating Tradition: Tourism Retailers in Guatemala. *Annals of Tourism Research*, 28(3), 658–687.

Pirsig, R. M. (1974). *Zen and the Art of Motorcycle Maintenance: An Inquiry into Values*. New York: William Morrow.

Tice, K. E. (1995). *Kuna Crafts, Gender, and the Global Economy*. Austin: University of Texas Press.

Trimble, J. L. (2006). Roe 'Duc-Man' Terry: A Chincoteague Island Gunner, Decoy Maker and Saltwater Cowboy. *Decoy Magazine*, (July/August), 40–44.

Urry, J. (2002). *The Tourist Gaze* (2nd ed.). London: Sage.

Chapter 8
Welcome to Wherever: Art in an Airport

Jo Ankor

Introduction

The proposal to have art in the new international airport terminal in Adelaide, South Australia, drew unexpected resistance from the developers of the new terminal. The particular art form had not yet been chosen; it was the placement of any kind of art intended to be representational of South Australia inside the space of the new terminal that was consistently refused during the planning, design and building stages. Indeed representational art that would portray aspects of the destination within this airport, as the new gateway to South Australia, was something that was consistently resisted over the several years of development and design prior to its opening in 2005. This was despite a range of submissions from the state Tourism and Arts authorities for displays, cultural exhibits, photographs, posters or even signage and interpretation of the site.

What was the basis of this resistance? Surely the international gateway to a large area of Australia should be pleased to announce its place, its people, its opportunities? Surely representational texts of some form would be expected to be integral to this space, to pronounce its identity and emplacement? What would the presence of representational art in a public space mean in the use and management of that space? These questions are explored in this chapter, starting with the limitations that the developers of the airport imposed on the submission for a set of images on existing glass panels within the terminal's design that was eventually negotiated only three months before the new terminal was due to open, and leading to an examination of the role of images in tourism and identity, the debates and issues of public art in public spaces and what these may mean for the tourist. To begin to explore this intersection of art and tourism, a critical tourism methodology which 'consciously avoids the empirical on the grounds that it may restrict the field of vision to only what exists' (Tribe, 2002, p. 340) is employed, meaning that the discussions in this chapter, although focusing on the empirical case of Adelaide Airport, are largely of a theoretical nature.

In terms of its structure, this chapter first briefly comments on the relationships between representation, art and airport spaces, then it establishes the context of the artwork proposal in the new Adelaide airport space as well as the nature of representational texts in relation to tourism, before moving on to examine the issues of art in public places that may influence tourists and tourism practices. A reflective summary considers the question of the 'transgressive' tourist when engaging in interpretation and exploration of the identity of place.

Representation, Art and Airport Spaces

Tourism research has identified strong links between representational images, motivation, choice, engagement and satisfaction in travel (see for example Butler and Hall, 1998; Gaffey, 2004; Garlick, 2002; Jenkins, 2003; Leisen, 2001; Nichols, 1981; Rojek, 1997; Selby and Morgan, 1996; Tuochino and Pitkänen, 2004). Representational texts provide sites for confirmation and the creation of meaning within the travel experience. As the airport is often the major gateway to the anticipated travel experience, not only giving entry into the new landscapes the traveller has chosen to access but also the point of departure and the final space of engagement with the chosen destination, this chapter suggests that representation of the host culture and landscape is critical in this liminal, gateway space. However, representational art is a statement of cultural and social identity, although not always an agreed one. Debates around public art as visual, representational portrayals of identity in a public space include arguments on the content, its interpretation and associated risks of provoking reactions, such as groups of people clustering around the artwork, or even people visiting that space just to see an artwork. In public spaces, playing with identity can be taken to be subversion and unexpected movement can be seen as a threat, creating tension with issues of security and control. These issues underlie the resistance of the airport developers to art in the Adelaide airport space. Yet, being a tourist is a form of experimenting with meaning and identity through moving into new spaces. The outcome then of 14 panels of glowing illustrations of aspects of the South Australian landscape and culture, installed in the public space of the Adelaide airport, is the radical giving of art to all users of the airport; travellers, tourists and operators alike. Public art is a visual practice that integrates, represents and communicates vision, image and space (Zebracki, Van Der Vaart, and Van Aalst, 2010). If art is 'an impulse to explore, understand and express ourselves, about ourselves and the world' as Garoian and Matthews (1996, p. 193) define it, there are similarities between art and the practice of tourism. The reflection on the meanings and purpose of public art which are central to the situation under discussion, require locating tourism and the tourist into a bigger picture, understanding aspects of the phenomenon of tourism in its placement straddling commercial business and aesthetic interpretive paradigms. The ambiguous role of the airport, as gateway for both people and goods, is problematised by the widely differing economic and cultural interests that intersect in the airport site, a site that is pivotal in much tourism practice.

The Airport Proposal

Wherever situated, each airport is a gateway. People arrive in locations and often their first engagement with that location is when they step into the airport space. Their final engagement may well be leaving from that same space. It is the gateway point where the first impressions are sought on arriving (to confirm an identification of destination and choice) or the final impressions are gathered at departure (to confirm the landscape visited, both cultural and geographical).

Adelaide, the capital city of South Australia, has a new state-of-the-art airport terminal. Negotiations between the South Australian Tourism Commission (SATC) and the airport developer, Adelaide Airport Limited (AAL), on displaying the cultural and natural experience of South Australia within the new airport space was attempted over several years prior to its opening. A number of proposals were submitted by the SATC, but all were rejected. Finally, only a few months before the new terminal was due to be opened in 2005, qualified support was given for the preparation and submission of an idea using a series of photo-panels to present South Australia within the airport space. Significant limitations affected the content of the proposal, beginning with the new terminal's design which relates closely to the geographical layout of Adelaide. Stretching along a north-south axis, the Mount Lofty Ranges forms the eastern boundary of the Adelaide Plain. The beaches of Gulf St Vincent form the western boundary. Centred in the plain is a mile-square City precinct, surrounded by the wide parklands that are an Adelaide feature. Suburbs fill the plain outside the parklands with close to a million inhabitants. The airport is situated on the western side of the plain, near the beach. The long, narrow terminal is aligned north-south, with an eastward-facing wall of glass almost the entire length and height of the building. This encompasses a spectacular view across leafy suburbs and compact city high-rise to a panoramic vista of the Mount Lofty Ranges. The architectural design included 14 glass panels (approximately four metres square) which abut this window-wall and partition the long internal space into bays. Some of the bays contain seating as departure and waiting lounges, while others are the entry points of walkways bringing arriving passengers directly in from the aircraft.

The SATC's proposal was to display images of South Australia on these panels to highlight quintessential and outstanding aspects of South Australia (Department of Premier and Cabinet, 2005). Approval was conditional, with limitations imposed by the developers that the glass panels were to retain transparency, or at least translucency, so that the long visual unity of the interior of the building was retained; any photographic images would need to be produced on translucent film which would then be affixed to the glass, providing an image while not blocking the interior's sightlines and equally presented to people on either side of the panel. Any treatment placed on the glass panels was required to be maintainable, that is, resistant to the hard use a public space incurs. The specific colours of the logos of AAL's airline clients were not to be present, for example the particular red colour of Qantas airlines.

The SATC's intention for these panels was to reflect and showcase the unique qualities of South Australia, including the sophistication, creativity and welcoming approach that make South Australia distinctive (Department of Premier and Cabinet, 2005). However, these values did not reflect the priorities of the developer, whose stated considerations, in their Annual report at the time, were managing (unspecified) on-going threats to aviation security, environmental issues and more efficient check-in and processing (AAL, 2004). This form of operational, technical focus acknowledges no other purposes than those of its economic business function, it remains insulated from other discourses (Tribe, 2002), and knowledge that does not further such aims is not considered in its operating functions. Certainly, providing visual photographic imagery which represents the geographic and cultural landscape in which the airport is situated and to which it leads, at the initial point of engagement (the airport terminal), is not a neutral action, rather it is expected to be emotive, to produce a response. The consideration of an emotive site within the airport is something that discourses of operational management may find difficult to even begin to express. If the primary focus of the airport, as stated by AAL (2004), is concerned with economic and processing functions, this structures airport space as one where other meanings of travel are denied or ignored.

The Design

The airport space has a role in travelling as an intermediary zone; the frame of mind of travellers at the point of their journey when they inhabit this space provides, indeed imposes, a context for their interaction with visual images such as the panels. It is also important to recognise the interplay of power and control in an entry/exit space that may conflict with visitors' preconceived ideas about the destination. These ideas, usually formed at least in part from promotional images of the destination but also influenced by culture, knowledge and previous experience, contribute to the motivation to travel to this particular destination and, therefore, the context within which the visitor looks (gazes) at both the arrival space and any representational texts encountered there. The introduction of visual representations, whose meaning is accessed through a process of reading and interpretation, produces a site of non-economic, non-controllable interaction within the arrival space. The regulation of the airport gateway environment for solely economic ends limits the possible, and necessary, interactions in the presentation of cultural meaning as embodied in the panels. If we understand culture as structuring how people think, feel and act, providing meaning and the basis of identity, here, suddenly, is a cultural interaction, colliding with an economic agenda at a site where contestation may occur and where interpretation and responses are encouraged.

What makes these panels encourage interpretation is that not only have certain photographic images been juxtaposed but that across them, a design technique known as 'paper-cut' has been applied. In this, a shape has been 'cut' out from the photograph, superimposing a second element into the framed picture. In one panel, the cut-out of a kangaroo gazes in surprise at the massed cyclists of the Tour Down Under road race; in another, a skeletal dinosaur crosses an ancient red desert landscape; in a third, a school of fish 'swim' across the panel which depicts the protected Leafy Sea Dragon. The photographs are of places and activities that can be found in South Australia while the cut-outs add a whimsical interpretive element to, and of, the photographic content. These thoughtful presentations can make the observer smile and pause to consider the references, identify the place, question the relationships between the images, those that are both photographic and cut-out. Through the 'cut-outs' a child can peer through, or a glimpse of the rest of the terminal can be taken, establishing a disconcerting, subtle yet deliberate, play between reality and representation.

Representational Texts

Photographic imagery is a familiar medium for expression of cultural expectations and in tourism and travel these expectations are often projections of the landscapes and culture of the destination. Therefore visual images of the host locality in an airport invite reaction and interaction. To explore how this occurs, an understanding is required of how photographic images convey cultural messages and how those messages and their interpretable meanings may be accessed; that is, the process of the reading and interpretation of visual representation as a form of text. The context in which photographs are displayed also provides a frame in which one meaning can become more prominent than another.

Our attitude to the photographic image is linked to the development of the camera, in all its shapes and forms that developed out of the 'camera obscura', a device adopted by artists around the eighteenth century. The camera obscura allowed the artist to gaze at the subject to be painted within an imposed frame. It isolated the subject from the surrounding context and enabled it to be examined and positioned in such a way that particular details or relationships could be focussed or highlighted, without in any way altering the detail of the subject. Despite widespread understanding of, and familiarity with, the digital technology that now enables images to easily be manipulated, the reliance on images to communicate what is 'really out there' in the world, whether as events depicted through an iPhone broadcast of breaking news or as places shown in documentaries, tourism brochures and websites, remains a central part of our understanding.

The 'positivist certainty' which photographic images are often perceived to have can now be undermined even further by technology (Batchen, 1991). However, in what situations do viewers really disassociate the image from the expected representation of reality? There is a contradiction in photographic images: the photographs do not change whereas the places, people, flora and fauna represented in these do. In the case of Adelaide airport panels, the whales swim forever in one, and the grapes are eternally ready for the vintage in another. Yet the panels are also a vehicle through which visitors are also informed about what South Australia is 'really about'. Representational art in whatever medium, while presenting recognisable figures or objects, is only representing the surfaces of visible reality and, in the case of photographs, this is done in a two-dimensional display of textures and light (Nichols, 1981). But photographs function because they express or communicate ideas beyond those recognisable surfaces (Frow and Morris, 1996).

Tourism images present places for consumption (Urry, 1990; Craik, 1991; Gaffey, 2004); art presents ideas for consideration. In presenting aspects of South Australia, the panels are an invitation to seek and engage with the images they show. Although people openly articulate the subject presented, it is more than simply the subject to which they are responding. There is an association of concepts that also occurs; for example, there may be an association between the rugged hills pictured and Indigenous culture, or with 'untouched nature', or perhaps with intrepid adventuring skills. The place pictured, then, is felt to offer a chance to experience another culture, or the natural world, or a skill. Whatever the associated concepts may be, it is this generated relationship between viewer, subject and concept that is important in tourism and in art. The response to the conceptual messages associated with the subject of the image is not often articulated but drives the motivation to travel and visit one place over another.

Art in Place

In the terminal of an airport, many images will be found such as logos, advertising, directions, however the inclusion of specifically artistic representations of the place to which this terminal opens are a conscious attempt to provoke a response in the users of the space. The expectation is that the place, once accessed, will be enjoyed and that these representations also are enjoyable.

A positioned object 'organises space, transforming it into place' (Tuan, 1980, p. 6) and the installation of artworks into public spaces transforms them by creating sites which offer an interpretation of the space in some form, sites where meaning is generated. Expectations of the use of the terminal space vary between the developers, the security personnel, the airline staff, the people waiting to meet and greet and the tourists, who all hold different and potentially competing expectations. Their presence in the space intersects but their needs and priorities do not. Public artistic expressions such as the panels serve to communicate the character of a given space, its history, its present and its likely future, they are 'spatialized processes of performing and displaying difference' (Zukin and Braslow, 2011, p. 138). The presence of a piece of public art endorses discussion in the realm of 'abstractness, motion, imagination and constant examination' (Dekel, 2009, p. 72). The art must be realised as also producing and disseminating aesthetic codes and symbols (Zukin and Braslow, 2011). While it is possible to walk away from a work in a gallery, once works are incorporated into lived spaces they cannot always be avoided (Sharp, 2007); the multivocality of the public artscape establishes the potential appropriation of these images by public space's users (Zebracki et al., 2010).

The high ceilings, imposing entrance and expanse of floor of the new terminal shrink the visitor into smallness, a similar relationship to that intended by the designers of Gothic cathedrals. Then, the idea was to spatially state people's relationship with God, that of humility, obedience, being in the presence of the all-knowing (a somewhat uncomfortable comparison). In the airport terminal, the observant, uniformed security guards, officials and airline personnel impose scrutiny, shrinking the visitor and directing interaction with both the space and the carefully selected activities offered.

Intersections in Place

According to Ward (1998), the role of public art is to explore representation and meaning of the space in which it has been placed. The first public art in Australia were the rock paintings of the Indigenous Australians whose relationship and responsibility with place is expressed in these paintings. The Australian Aboriginal people understand that the act of marking the landscape changes the relationship between place and person. An exploration of place occurs within the panel content, as it is specifically designed to do, though meaning is determined by individual interpretation. The tourist's response to place, the knowledge of place they bring with them, will be reinforced, perhaps augmented, possibly challenged, by the pictorial information of the panels.

The information supplied within the airport space plays a critical role in the visitor's experience at that vital point of first impression. For example, the viewer receives information from one panel that in South Australia there is a popular sporting ground where cricket is played. For cricketing specialists, knowing that the pictured oval is a world-famous cricket venue and that Adelaide was the final home of famous cricketer Sir Donald Bradman would add to their response to the panel. Tourists bring their cultural knowledge into the destination space, with the intention of experiencing that place. The tourist arriving in Adelaide to attend a cricket event would have their expectation of place confirmed by the panel that depicts this activity.

As mentioned earlier, the idea of place held by individuals is, among others, also influenced by images circulating in their culture (Butler and Hall, 1998). This interaction of image and expectation has been described as the intersection of imagination with reality (Rigg and Ritchie, 2002 in Gaffey, 2004). The success of this connective intersection is reliant on there being a shared knowledge between producer and receiver, on both the producer of the image, whether poster, brochure or panel, and the viewer of that image, having some shared cultural knowledge. Recognising the connection constructs both the identity of the place being represented in the text and the reader's identity as one who understands subject and concept. The way a photographic image invites the viewer to look into it, at its presented surfaces of reality, serves to position the viewer in a particular way with the subject of the photograph. This relativity of perception therefore also places the viewer in a relationship with the meanings available. The tourist is part of a complex interaction with image and intention and the accessing of meanings in signs. According to Urry (1990), the postmodern tourist works constantly, playfully and rapidly through a pastiche of experience and meaning. Tourism practitioners work to establish tourism income from mutually-enjoyable, renewable, transaction processes. The difficulty is that these are experienced through feelings, perceptions and attitudes; through learning, education, fun, engagement; through concepts such as relaxation, escape and search. The constant tension in tourism as an industry is between the seeking and consumption of intangibles that are numerous, subtle, responsive and descriptive interfaces and economic paradigms and processes. In this industry, the economic and cultural come face-to-face, inextricably intertwined and interdependent. The airport is a powerful example: an intersection zone of arrival, departure and discovery but also of motivational and economic priorities, movement across boundaries both physical and ideological.

Liminal Movement

In tourism, the movement through, or crossing of, physical boundaries is matched by the crossing of cultural boundaries where the visitor enters with the tools of their own habitus and culture to hand. In the airport, different scales of mobility intersect, from global movements of international travel to the local terminal activity (Adey, 2004), resulting in a mixture of *foci* and expectations that is constantly changing. The airport's position as a constantly changing space of intersection between entry and exit renders it liminal in nature. A liminal zone is defined by its location as a space which is different from those which lie adjacent to it, flanked by separate and irreconcilable alternatives, a space between. The intertidal zone of a beach is liminal; at times a water-covered ocean environment, at other times a sunny land environment. It is also in constant change and predictability is compromised because rules cannot be established that consistently apply; it resists definition. Because of the fluid nature of the liminal space, however, each participant brings their own belief of what constitutes appropriate behaviour within that space. The rules that each establish are not always recognised between one group of users and another. Indeed, Hollinshead describes the liminal space as 'garbled in signification and filled with temporal restlessness' (2004, p. 35).

When such places are the borders between nations, then issues of governance of activities in these zones and the control of behaviours, not to mention surveillance of users, are imposed on a space already problematic in definition. Travel is a specific kind of liminal space characterised by certain temporal rituals of leaving, moving and returning (Fullagar, 2002) and the airport is a place occupied between leaving and arriving, poised between planning and experiencing. As a liminal space, it carries not only the acknowledgement of border but also the perception of suspension of rules, a sense of being in-between the corporeal and rational realms of experience (Fullagar, 2002); the traveller has left predictability behind and is not yet emplaced in the new. Being in the border or gateway zone, where one is neither in nor out and having left, one has not yet arrived, has an impact on the thoughts and attitudes of those travellers present.

The excitement of the liminal place infects and affects the airport. It is an exotic environment because of its very liminality, its sense of being exclusive and outside the ordinary. Without experience and knowledge, the visitor can quickly be intimidated (a feeling of not belonging, of confined to a place where rules and identities are confusingly mixed) and bewildered (the risk of transgression and consequent rejection). For, in fact, rules abound. If tourist users of the airport are already in the disposition of being outside the rules of home, for they are on the journey to experience difference and to try the new, how do they know what rules may be asserted? And if an interpretable site of representational art exists within this liminal zone, a site which invites plurality of reading and response, what does this mean for the rule-makers, the rule-enforcers? Carter (2005) identifies a lack, or deliberate effacement, of identity in contemporary public spaces because it is a source of difference and debate. The provision of functionality without personality, of surveillance that disallows being, of faceless space, is established to counteract the potential for subversion of interpretation, for any non-anticipated generation of ideas.

In the postmodern world, collaged identities, multiple realities and constant change are accepted and, indeed, expected. But at the rigid boundary of nation-states, the playfulness exhibited by the postmodern position will emerge as subversion, be it subversion of space or subversion of the authorised photographic representations contained in the panels. The artwork may not necessarily propose alternative perspectives 'but opens up an intellectual and sometimes literal space for this possibility' (Richardson, 2010, p. 19). In presenting the panels in a public space, disparate readings are inevitable. Images such as those used in the panels convey powerful messages of identity. When tourists arriving at the airport view the panels, they are interacting with a text; they are establishing the points of difference, spatially and culturally, between themselves and the host destination to construct a personal and social identity within what can be described as a temporary interpretive community (Fiske, Hodge, and Turner, 1987). Art in this context leads the viewer to experience the world through different sensibilities, and, at its deepest level it can challenge its viewers' belief system and affect the way they engage with the world around them (Bayles and Orland, 1993).

Art, Tourism and Transgression

Organised systems, such as airports, 'produce willing social participation appropriate to their presumed and stated objectives' (Richardson, 2010, p. 19). When mobility studies use terms such as nomadism, diaspora and displacement (Adey, 2004), the concept of mobility, of moving through places, carries a variety of transgressive connotations, escape from social order, an act of resistance. Where does the tourist fit?

Movement of the body often signifies the transgression of social norms and is therefore viewed in terms of risk to 'the safe and static containers of space, territory and social order' (Adey, 2004, p. 502). The term 'container' itself denotes established areas of inside and outside. Bauman's (1996) identities for the postmodern traveller include the pilgrim, the stroller, and the vagabond; figures who perturb, who resist being contained. Visitors with cultural knowledge acquired in their own social or cultural setting, therefore, may transgress the boundaries of ideological and cultural space as they cross physical borders to arrive in the destination. The line between sanctioned tourism and a structuring of tourists as transgressive is as fluid and indefinable as in any liminal zone. What is required, therefore, is a certain legitimisation of transgression, an allowance for the incorrect or unpredicted, within the airport space.

An artistic intervention can potentially create confusion, the resolution of which requires an implicit negotiation among those who share the space (Richardson, 2010). In the Adelaide airport terminal, the control of moving bodies through the space, the surveillance of movement, the desire for conformity over chaos that airport management wants are positioned against a set of images which require exploration. The panels are texts that are produced within and by the culture of South Australia; they are deliberately intended to introduce as many aspects of South Australia's culture, in Williams's (1989) broad sense of culture as the everyday expression of living, as possible. They represent the place that is on display outside the terminal's large window-wall.

Representational texts are not literal but contain associative concepts and meanings. Often there is a gap between what the intended meaning, or a range of meanings, might have been and the new meanings that are generated through the interaction of the text and the cultural skills of the individual reading that text. The establishing body's expectations of both accessed meaning and response to an installation of public art are often strongly defended when faced with the public's alternative readings. Foucault's (1980) concept of a truth dictated by political and economic interests encourages the questioning of such presentations to uncover their underlying ideologies. And in the airport site, two conflicting ideologies, that of economic rationalism and that of the aesthetic in the generation of cultural exchange and interaction, are present in an uneasy association of differentiated intention. Inserting representational, interpretable texts into such spaces adds further complexity and may become a clearly definable site where differences are focussed. The airport lounges that tourists flow through are initial sites for the seeking of difference that is part of travel (Robinson and Phipps, 2003). This is because travel is not so much an escape from everyday life as an activity that provides a plane of cultural difference in which everyday life routines are contrasted and developed (Rojek, 1997) and against which identity can be re-tried and established. Interpretable cultural texts placed into an airport will be accessed by some travellers as a confirmation of motivation to come to this destination, and as potential sources for those moments that take the tourism experience to an intensity that validates the work involved in travelling.

Conclusion

In this chapter, a critical approach has helped take us beyond the act of installing an artwork to a consideration of the nature of the space in which it stands, of representational art and of responses to art, all within a tourism context. It has raised questions about tourists' engagement with public art. In doing so, it uncovered issues of the nexus in tourism of place, person and culture, and the engagement with art in public places, that have not been considered together previously in tourism studies.

The complexity of the interactions between tourists, culture, representational art and art in public places cannot be ignored. While the panel installation is present at Adelaide airport, the subsequent arrangement of television monitors, seat orientation, boarding counters and the like, has rendered some panels as background. Yet the art remains and the airport functions. The combination of public art and liminal public space, of interpretable sites in a borderline environment, entails accepting a level of risk as far as the interpretation of representational art is concerned which is at odds with the control of airport space. It is this ambiguity that underlies the tensions of the panel installation and one of the constant tensions embedded in tourism, the host's need for control and the tourist's need for intangible experience. This chapter proposes that risk must be accepted and that representation of the host landscape is critical in the liminal/gateway space. Understanding these interactions establishes the possibility of a discourse accessible to both economic and cultural perspectives. Hopefully, the panels will continue to contribute to a greater enjoyment of being in South Australia for all travellers passing through the airport terminal. Without these, it would certainly be an entry to: what is the name, now?

References

Adelaide Airport Limited (AAL). (2004). *Adelaide Airport Annual Report 2004*. Adelaide: AAL.

Adey, P. (2004). Secured and Sorted Mobilities: Examples from the Airport. *Surveillance & Society*, 1(4), 500–519.

Batchen, G. (1991). Desiring Production Itself: Notes on the Invention of Photography, in R. Diprose and R. Ferrell (eds), *Cartographies*. Sydney: Allen & Unwin.

Bauman, Z. (1996). From Pilgrim to Tourist – Or a Short History of Identity, in S. Hall (ed.), *Questions of Cultural Identity* (18–36). London: Sage.

Bayles, D. and Orland, T. (1993). *Art & Fear: Observations on the Perils (and Rewards) of Artmaking*. Santa Cruz: Image Continuum Press.

Butler, R. and Hall, C. M. (1998). Image and Re-Imaging of Rural Areas, in R. Butler, C. M. Hall, and J. Jenkins (eds), *Tourism and Recreation in Rural Areas* (115–122). Chichester: John Wiley & Sons.

Carter, P. (2005). Writing Public Space: Design, Philosophy, Art. Paper presented at the Research Seminar Retrieved 19 September 2005, from <www.londonconsortium.com>.

Craik, J. (1991). *Resorting to Tourism*. Sydney: Allen & Unwin.

Dekel, I. (2009). Ways of Looking: Observation and Transformation at the Holocaust Memorial. *Berlin Memory Studies*, 2(1), 71–86.

Department of Premier and Cabinet (2005). *Adelaide Airport Cultural Panels – Designer's Brief.* Brief prepared by J Layther and J Montessi. Adelaide: Department of Premier and Cabinet.

Fiske, J., Hodge, B., and Turner, G. (1987). *Myths of Oz: Reading Australian Popular Culture*. Sydney: Allen & Unwin.

Foucault, M. (1980). *Power/Knowledge: Selected Interviews and Other Writings* (C. Gordon, Trans.). Brighton: Harvester Press.

Frow, J. and Morris, M. (1996). Australian Cultural Studies, in J. Storey (ed.), *What is Cultural Studies? A Reader* (344–368). London: Arnold.

Fullagar, S. (2002). Narratives of Travel: Desire and the Movement of Feminine Subjectivity. *Leisure Studies*, 21, 57–74.

Gaffey, S. (2004). *Signifying Place: the Semiotic Realisation of Place in Irish Product Marketing*. Aldershot: Ashgate.

Garlick, S. (2002). Revealing the Unseen: Tourism, Art and Photography. *Cultural Studies Review*, 16(2), 289–305.

Garoian, C. and Mathews, J. (1996). A Common Impulse in Art and Science. *Leonardo*, 23(2/3), 209–214.

Hollinshead, K. (2004). Tourism and New Sense: Worldmaking and the Enunciative Value of Tourism, in C. M. Hall and H. Tucker (eds), *Tourism and Postcolonialism: Contested Discourses, Identities and Representations*. (25–42) London: Routledge.

Jenkins, O. (2003). Photography and Travel Brochures: The Circle of Representation. *Tourism Geographies*, 5(3), 305–328.

Leisen, B. (2001). Image Segmentation: The Case of a Tourism Destination. *The Journal of Services Marketing*, 15(1), 49–65.

Nichols, B. (1981). *Ideology and the Image*. Bloomington: Indiana University Press.

Richardson, J. (2010). Interventionist Art Education: Contingent Communities. *Social Dialogue, and Public Collaboration Studies in Art Education*, 52(1), 18–33.

Robinson, H. and Phipps, A. (2003). Worlds Passing By: Journeys of Culture and Cultural Journeys. *Journal of Tourism and Cultural Change*, 1(1), 1–10.

Rojek, C. (1997). Indexing, Dragging and Social Construction, in C. Rojek and J. Urry (eds), *Touring Cultures: Transformation of Travel and Theory* (52–74). London: Routledge.

Selby, M. and Morgan, N. (1996). Reconstructing Place Image. *Tourism Management*, 17(4), 287–294.

Sharp, J. (2007). The Life and Death of Five Spaces: Public Art and Community Regeneration in Glasgow. *Cultural Geographies*, 14, 274–292.

Tribe, J. (2002). The Philosophic Practitioner. *Annals of Tourism Research*, 29(2), 338–357.

Tuan, Y. (1980). Rootedness versus Sense Of Place. *Landscape*, 24(1), 1–8.

Tuochino, A. and Pitkänen, K. (2004). The Transformation of a Neutral Lake Landscape into a Meaningful Experience – Interpreting Tourist Photos. *Journal of Tourism and Cultural Change*, 2(2), 77–93.

Urry, J. (1990). *Tourist Gaze*. London: Sage.

Ward, S. V. (1998). *Selling Places: The Marketing and Promotion of Towns and Cities, 1850–2000*. London: Routledge.
Williams, R. (1989). *Resources of Hope: Culture, Democracy, Socialism*. London: Verso.
Zebracki, M., Van Der Vaart, R., and Van Aalst, I. (2010). Deconstructing Public Artopia: Situating Public-Art Claims Within Practice. *Geoforum*, 4(1), 786–795.
Zukin, S. and Braslow, L. (2011). The Life Cycle of New York's Creative Districts: Reflections on the Unanticipated Consequences of Unplanned Cultural Zones. *City, Culture and Society*, 2, 131–140.

PART III
Travelling Artists

Chapter 9

The Nomadic Village: Communal Creativity and Political Subversion in a Temporary Settlement

Sharon Wilson and Pau Obrador

Introduction

This chapter is concerned with nomadic artistic practices and the transformative and creative qualities of tourism. We are interested in how diasporic artists find meaning and context in which to produce art in temporary settlements outside the traditional gallery environment, and by doing so how these diasporic figures oscillate between the figures of homelessness, estrangement and belonging. This chapter looks at the interplay of travel, tourism and art by examining the *Nomadic Village*, an Arts Council funded project that took place in May 2012 in County Durham, United Kingdom. The site of enquiry consisted of an assembly of mobile living spaces within the rural village of Wolsingham. It was occupied by 30 artists from countries including the United Kingdom, the Philippines, Austria, France, the Netherlands and Australia who used the natural setting as a space to live, work and play. This chapter, thus, seeks to comprehend these itinerant hypermobile bodies who engage in a performance of expressive individualism born out of shared values of autonomy, experimentation and political resistance.

This chapter raises three critical issues concerning the interplay of travel and art. Firstly, there is the issue of creative nomadism. The case of the *Nomadic Village* brings to the fore how the human compulsion to produce art leads to the formation of 'transnational migrant circuits' (Rouse, 1988, p. 26). Diasporic artists have a lot in common with traditional forms of nomadism (Weissleder, 1978) moving from place to place rather than settling in one location. However artistic nomadic practices are not borne out of a necessity, but are a lifestyle, the product of a learned appreciation. D'Andrea (2007) has introduced the concept of neo-nomadism to examine these global circuits of counter cultural practices that unfolds expressive forms of individualism. By understanding ways in which these elements converge, insights are offered into the modes in which art based roving lifestyles can articulate the social and cultural practices of tourism today.

The second issue is the association of travel and creativity. The case of the *Nomadic Village* draws attention to the artistic dimension of tourism itself (Obrador and Carter, 2010). In the project, travelling was an object of artistic endeavour as well as a way of looking around, a lens artists use to produce new intensities. This chapter offers insight into the modes in which art is produced in response to mobility and communality. By linking creativity and tourism, the case of the *Nomadic Village* gives credence to the idea of 'exile as the vocational imperative' (Levin, 1966, p. 38). Such an idea is central to contemporary culture. Modernist and post-modernist tropes have recurrently emphasised the aesthetic and intellectual benefits of exile and detachment (Kaplan, 1996). Absolved from the material conditions and power relations that make it possible, tourism is presented as a space of intellectual work and creativity. However creative forms of nomadism are not socially neutral, but only available to the fortunate few. According to Kaplan (1996) exile places upwardly mobile travellers in a position of 'exclusivity'. This chapter seeks to find out whether power relations between the voyeur and their subjects were reinforced due to the 'extraterritorial' nature of the global nomads.

The final issue the *Nomadic Village* project raises is political. Historically many of these communities have been perceived as 'vagrants' and have often been subjected to persecution in the guise of containment, control, assimilation and dispersal (Okely, 1983). In contrast the creative nomads of this chapter were welcome. They benefited from a public funding grant and a formal invite to stay in Wolsingham for 10 days, although not without initial reservations by the local council. This situation raises questions around local resistance as well the performance of hosting and guesting. It also draws attention to a dichotomy between political-economic institutions defining contexts of constraint and deterritorialization, and expressive tourists subverting dominant uses and definition of space. In so doing, the *Nomadic Village* echoes the ideas and practices of the situationalist avant-garde (Pinder, 2005) who called attention to the potential of art to disrupt the geographies of the everyday, advocating for the transformation of spectators into creative actors. Social space in turn can be socially, culturally and politically reconfigured as a new social order

Whilst this discussion uses a theoretical underpinning to look at issues around artists being away from their birthplace, seeking to find social freedoms whilst engaging in creativity, this chapter is informed by the fieldwork data collected in a 'nomadic ethnography' (as in D'Andrea, 2007). Our methodological approach was empirical. In order to collect the data, one of the researchers taking an immersive approach lived alongside the respondents in a 1972 VW Campervan and participated in community activities. Using the vehicle as an interview space core material was collected using a semi-structured interview technique with 25 audio recordings taken, transcribed and coded. Additional materials were generated through a research diary kept for observational work, photographic evidence and note taking of informal conversations. Because one of the researchers was also a visual artist, they further integrated into nomadic life by helping other artists with their work. According to Lett (1990, p. 131) 'the insider can learn to analyse like the outsider' thus the final etic account can have an emic origin. Living in the village as an artist meant that we had to be mindful of our own subjectivities, and of the fact that one of us was a visual artist and therefore in a powerful position of studying their 'own' (see also discussions in Rabe, 2003). that objectivist approaches can be blurred by researchers studying their 'own'. The main questions asked sought to confirm social backgrounds, propensity to travel to make art, personal interpretations of the *Nomadic Village* and finally to explore the connections or disconnections between artistic nomads and the host village of Wolsingham. While grounding the discussion within the qualitative empirical data produced, the resultant discussions aimed to address the key themes within tourism theory drawn from the literature review.

Neo-nomadic Practices

Photographer and self confessed artistic activist Klaus Mahering, like many other contemporary nomads was brought up in a middle class background. In 2007 he left the advertising industry with a view to obtaining a more tranquil lifestyle by converting a Steyr-Ikarus bus into an atelier-office-and gallery-space for the purposes of living a nomadic existence (see Figure 9.1). Despite owning an apartment in Austria he decided to take to the road, live on limited resources and to produce work not for deadlines, but only when he was inspired to do so. Moving forward, Klaus began to place more emphasis on combining his professional interest in creative practices with his travel aspiration and the spirit of communality by inviting friends

Figure 9.1 Klaus Mahering, the captain of the nomadic village

Source: Photo by Karim Blanc and Peter Westman, used with permission.

and colleagues to join him in building an outdoor workplace where multi-media installations and happenings could develop. Having been involved in other travel based arts initiatives before he developed his first *Nomadic Village* in Bulgaria in 2009 (On-The-Road-Productions, ATRP). In 2012 he organised a second edition of the *Nomadic Village* this time in County Durham, United Kingdom. Being on the road meant being somewhere else, in between things that hadn't been pigeonholed yet. In the second edition a greater number of invited artists stepped out of the sedentarism of everyday life to create artifacts about their experience of dislocation. They brought their own housings such as buses, vans or caravans, which became the structure of the temporary village. Many of the participants had been involved in similar projects before, with some even living a squatting lifestyle.

The *Nomadic Village* was a temporary community of artists not borne out of an economic necessity, but it is the result of conferring a special creative meaning to travelling. It is a paradigmatic example of a hypermobile fluid culture in which spatial displacement is intertwined with new forms of identity and subjectivity. Anthropological studies of nomadism provide a conceptual foundation for understanding such contemporary hypermobile cultures (D'Andrea, 2007). The *Nomadic Village* is more than a simple expression of cosmopolitanism, networking or globalisation. Like in the case of pastoral tribesmen, it is a cultural formation that establishes a deterratorialised relation with space, thus defying a sedentary logic of art. It is a community in which there is an evasion of dominant sedentary apparatuses and travelling is a source of learning. However there are also important differences. Classical nomadic cultures such as the Massai and the Turkana have formulated nomadic living patterns out of a critical need for survival. It is crucial for them to travel together to pool resources and to maintain a degree of social harmony. However in the case of the *Nomadic Village*, nomadism has nothing to do with survival but it is ultimately a lifestyle choice. The travelling artists like other expressive expatriates were not driven by financial rewards; instead they were looking to share their ideas of freedom and creativity with others with similar ideals and backgrounds. The travelling artists deploy the tactic of being on the move with a creative purpose, thus blurring the classical distinctions between art and leisure.

The *Nomadic Village* is a project that combines mobility with self-imposed marginalisation. This is a group that to some degree abandon the safety of home to co-inhabit the fringes of modernity and society by reconnecting with a commune. An anti-establishment theme was evident throughout the project. Michael from the Philippines described life at the *Nomadic Village* as an alternative community where everyone is caring, compassionate with one another: 'We start as strangers but when we come together to eat it changes I think. There are no boundaries between people, it's just one establishment'. For Michael there was a sharp contrast with mainstream society where he described modern life as, 'People having to act like robots'. Whilst conversely he felt inside the village was different. For example everyone ate together at the same time which was something that he felt happened less and less except for in prisons and hospitals.

Contemporary nomadism is a counter cultural lifestyle that rejects a utilitarian view of modernity in the name of freedom and creativity. Spatial displacement here is a means of escaping the sedentary logic of the arts. Marginalisation is a fundamental principal of contemporary hypermobile cultures (D'Andrea, 2007). Exile is supposed to provide aesthetic and intellectual gains, opening a field of creativity that is not available under the conditions of hegemonic praxis. Whilst creative nomads may have rejected the sedentary logic of the state by being mobile (Briadotti, 1994), a utopian idealism was difficult to sustain, only possible by artists going from one funded arts project to the next.

The idea of the village appeared subject to mixed appropriation by the different stakeholders such as funders, project managers, visitors and participants. The *Nomadic Village* was interpreted as everything from a social laboratory, a corporate networking opportunity, a passive political demonstration, an art school, and a working holiday and as a place of leisure where nature and communality seemed to be enjoyed. Sometimes conflicting ideas about what attending the event could offer was expressed by artists. Some used the opportunity to career build as one might network at an International conference, informally exploiting 'opportunistic' pathways of social mobility. Others responded to the idea of living together with un-familiars to build professional friendships leading towards creative partnerships that they had not yet imagined. Katie from London commented that the facilities were better than her house; whereas Boris from the Netherlands (the only designer in the camp) was surprised that all human needs were catered for. In his view a greater reliance on each other and sense of direction would have grown more profoundly from the experience of living as traditional nomads. In addition, Michael from the Philippines, for example, said: 'I could not take photographs of the English countryside without being in it, thus living here has provided inspiration to give birth to art, not otherwise possible', explaining in such a way that just having the *Nomadic Village* as a place to make art in another country was valuable in itself.

While emphasising marginalisation, the *Nomadic Village* was a highly selective community of nomads. The choice of artists was not accidental but socially engineered. Participants were selected according to a set of curatorial paradigms to facilitate artistic creativity, social cohesion, encourage global networking and knowledge transfer. The international cohort of participants came from places such as France, New Zealand, Austria, the Philippines and Germany with the remaining third from the UK. Some participants had not met before, whilst half had completed the first *Nomadic Village* project thus familiar with Klaus. Social bonding of the community relied on a belief that meaningful activity arranged around creative labour was conducive with a feeling of freedom and self-actualisation. Michael in describing being a nomad as 'living the dream' also echoed this sentiment. He said that sitting in Klaus's bus with other artists felt like family forming. However not everyone agreed on the idea that art had the power to hold people together. Stevie, a local artist from Northumberland, for example pointed out that even the most visionary plan of social engineering might not induce the expected results. He did not think that an application process would always lead to a good outcome, believing that it was often down to good luck how people socialised. A strong functioning community Stevie maintained was still quite fragile, as 'it would only take one person to be very critical of other people to upset things. I have participated in residencies before with a clash of personalities which rendered them unworkable'.

The community of artists that was assembled in Wolsingham corresponds with Maffesoli (1996) and his description of neo-tribalism. This was a fleeting community of strangers bound together not so much by an ideology or a political apparatuses as by a lifestyle or an emotion, in particular the need for a more expressive life. Like in the case of Maffesoli's neo tribes what matters here is the being together as well as the creation for its own sake. Giddens (1991) suggests that these new social networks define a 'post-traditional order' in which diversity finds some organisation although is highly disjunctive and lacks anchorage. Such postmodern forms of sociality challenge the distinctions between pleasurable work and a proactive vacation, breaking with a clear-cut division between work and leisure, home and away. In contrast to modernised cultures of industrialisation where normative working practices are often regimented, time-based and venue-related, the *Nomadic Village* promoted a lifestyle that brought work and leisure together. Unlike regular employment there was no enforcement for nomadic artists to produce particular outcomes or to socialise in specific ways. For example there were no 'icebreaker' workshops, team briefings and no working structure outlined by ISIS Arts Management, just informal gatherings that no one was obliged to turn up to. So whilst the *Nomadic Village* was a workplace which accommodated office desks, laptops, photography labs and mobile cinema spaces, if people engaged in leisure activities instead, no value judgements were made about them not 'working' in the traditional sense. In fact Margaret an installation artist did plan to start making videos as soon as she arrived, yet was surprised to discover that having conversations, eating together and spending leisure time with others also felt like productive time, and, for some of the dwellers it often took precedent.

The *Nomadic Village* was an event that was positioned in a public picnic area in the village of Wolsingham. As with the choice of artists, the physical design of the village was also not accidental. From the arrival of the nomads onsite, Klaus began to organise how the village would look, encouraging the vehicles to form as a circular shape to create a space where communality would be easier. With vehicle exit doors facing inwards, all the residents could face each other. This aesthetic arrangement however also constructed a wall boundary between the artists and the public, defining an enclavic space with a potential psychic boundary. None of the artists thought that the fence made of mobile homes was a conscious plan, but one which seemed to happen naturally as the vehicles arrived. In other words they did not think that Klaus was being strategic about how the design would work as social construct, but instead surmised that as he was an artist he needed to create a sense of aesthetic order. The large open central area was used for communal yoga, ball games, performance and some social gatherings. Artists tended to congregate in the communal marquee placed at the end of the village next to the project management area. This hub was set up so participants could access wifi and create ad-hoc social events. Its main function however was to use it as a dining area, a ritual that was signified three times a day by the sound of someone tapping a pan with a stick.

Nomadic Art Practices

The *Nomadic Village* offered a creative platform for 30 nomadic artists from around the world, taking up residence transforming the itinerant enclave and surrounding woodland of Wolsingham from a tabula rasa, devoid of 'art' into an open air showcase of diverse contemporary artistic practices. By exploring the nature of nomadic art thus, this section provides an overview of the work produced as a way of exploring in this context, the inextricable ties between art and travel. The geographical displacement of their creators had made possible a multi-disciplinary exposé of drawings, poetry, performance, film, documentary photography and various conceptual artworks contrived from objects collected by the nomads whilst in transit.

A significant part of the art produced at Wolsingham was made as a response to being mobile. Travelling was an artistic endeavour as well as a lens, which artists used to create new intensities in different places. The work of Caitlin and Andrew Webb-Ellis examples one of the many artistic practices that speaks directly about travelling and dislocation. They brought a 16mm film about the cultures of 'Extreme Running', which they intended to show at the Cannes film festival. Their interest in human mobility was a good fit with the residency for two reasons. Firstly the artists had travelled to Norway and Slovenia as well as a number of UK locations to follow the subject of their documentary. Secondly their theme human movement embodied the consequences of travel.

Margaret, an American performance artist living in Wales, also mediated links between travel and art via her low fidelity video work. She brought a film called 'Britain in a Day', which expressed her interest in how people consumed places as tourists. Margaret as self-styled quirky narrator, posed as an 'alternative' tour guide at the seaside town of Penzance. As the central character in this tragic comedy travelogue, the viewer is taken sightseeing on location and subjected to the daytrip going wrong. Mobility was also reflected in the work of local artist Stevie Ronnie who arrived in a converted transit van, onboard a mobile art library. He unpacked his collection of handmade art books and installed them on temporary tables outside the vehicle. The visiting public were then invited to produce their own masterpieces which would later be shown in other locations.

One exemplar of nomadic art in the village was the work of Alison from Yorkshire. She had been living a mobile lifestyle for the previous 18 months, choosing to leave a prestigious art gallery job. Her work illustrated how she travelled with her family in a campervan and the consequences of the fluidity of experiences generated by living in a continuous state of flux. Using her taxidermy skills she made what she described as shamanic art from dead animals found on the road. Most of the things she created went hand in hand with a self-imposed exile. Due to basic living standards in the van, for example having no electricity, meant that much of her work had a 'make-do-and mend' aesthetic quality. In other words 'limitations' on lifestyle due to non-fixity determined how things were done and thus her artworks were an expression of her mobility. To her mind one could not get closer to travel and art than picking something up that's been killed on the road, eating it and making something beautiful with its skin. Other works included herself and her partner performing the eating of her preserved placenta. They had brought it to the village with the intention of ritualising the special moment by sharing her birth processes with the nomadic tribe. The placenta skin was dried in the sun and made into a shamanistic drum afterwards. This intervention was followed by Alison taking the remaining blood, mixing it with black food colouring and applying it like ink to paper. Purposefully using non-sophisticated craft equipment, she tried to emulate the simplistic approaches to creativity used by ancient tribes, as a way to symbolically detach from the rationalised production values associated with modernity. To do this she repeatedly painted a stylised self-portrait based on a Sheela-Na-Gig, an archetypal pre-Christian image of fertility. She also liked the idea of leaving traces of herself behind before moving on from the residency. Alice's work did not only speak of travelling, it also emphasised a creative need for self-marginalization.

Many of the artists at the *Nomadic Village* sought alternative lifestyles in order to be creative. Take for example the case of Katie. On her way to the residency she had picked up debris such as old wood, broken doors, and discarded junk from the roadside and planned to sculpt with them. The fact that her artwork would be dismantled the day after it was completed, determined how it was made. Thus the art produced due to its nomadic disposition was not only the materiality of travel but was also destroyed by such conditions. Katie pointed out that her work echoed her current lifestyle of squatting in London and felt it had a more sublime quality because it was temporary. Like Alison, Katie used self-marginalisation as a way to be creative. Many of the artists who enjoyed alternative ways of living could be perhaps described as opportunistic 'lifestyle entrepreneurs' (Peters, Frehse, and Buhalis, 2009). Despite the idealistic attitude some of the artists exerted, most used the *Nomadic Village* as a career development opportunity. Most artists seemed reflexive and able to change direction quickly or compromise to some extent to exploit the situation. For example when Katie was asked if it was feasible to build her installation in an art gallery elsewhere, she explained that re-contextualisation of the artwork would be an option if she was financially reimbursed.

The *Nomadic Village* was as much about cooperation and community engagement as it was about nomadism. This is a community that chose to commute and collaborate which led to an eclectic mix of art produced. That said, artists used the opportunity in different ways. Some artists saw it as a place to relax, where they could be creative should they feel inspired. It was an opportunity to be away from home in the company of likeminded others. Other artists saw value in collaboration for its own sake and seized the opportunity to establish new cooperative forms of work. This was the case of poet Stevie who whilst his main concern was his mobile library, in the spirit of making transnational connections, he collaborated in a performance with Quinn, a dancer from Australia. In contrast some artists were less interested in public engagement and preferred to work alone. Boris an environmental designer from the Netherlands for example, because of the communal aspect of the project, felt obliged to share his work with the passing public. He provided engagement by allowing visitors inside his mobile living space to look around. At a 'show and tell' event however, Boris had no problem sharing with nomadic peers how his Silo pods were made or how he transformed old car tyres into modernistic furniture, but he rejected the use of corporate etiquette in favour of an informal chat about his work over a few beers.

While community orientated artists saw the process of making art with people with no art training as part of their practice, others whom normally made work on a solo mission required payment to engage creatively with the Wolsingham community. As Lippard points out artists are 'caught like prostitutes between a rock and a hard place' (1999, p. 77) finding themselves compromising their practice to earn a living. That said, we are not suggesting that all artists saw the idea of social engagement as an inconvenience, instead for some it was their strength. Maartje from Holland for example usually made her own films, but due to the incentive of payment was prepared to help teenagers from a nearby secondary school make a video. The macabre comedy re-enactment was set in a local graveyard and was played out by the young people as a spoof horror film. The work did not have high production values, but instead was much more about educating children on basic film skills and encouraging their interest in local history. By providing the opportunity to work with an artist from another country, hosts and guests were brought together by the *Nomadic Village* so networks between travel, tourism and art could be formed.

There were many connections and crossovers between leisure, tourism and creativity in the *Nomadic Village*. The most touristic project of all was Alan and Andrew's installation. They installed a pre-built Caravan Hotel onsite so that artists not chosen in the selection process could book in as visitors and stay for one night. Because the caravan interior was designed beforehand, the two artists felt no obligation to do anything during the residency other than greet the guests when they arrived. Allen took an interest in the foreign artists and interviewed some of them, in between socialising and networking, however the main work for them had already been done. This meant that the rest of their time could be spent as they said 'chilling out'.

Stevie's mobile library was another example of crossover between creativity and tourism. Passers-by were offered the chance to join in with a 'make your own art book' workshop and were encouraged by Stevie to produce small works that would be exhibited in other places when the bus went on tour. The main point here is that books were made by people who were not artists, yet their contributions in this context were seen to be just as valid by the artists and curators as those produced professionally. Perhaps the quality of the objects were not the main concern, but because they were due to be displayed in other places, it galvanised with them cultural capital. Stevie had also brought suitcases with him which Klaus suggested be given to the artists so they could make work to put inside them. Artists customised suitcases by filling them with objects found on location. The researcher also made a clay model of the *Nomadic Village* and added an MP3 player which produced an audio recording of the natural sounds, whilst others included photo archives and selections of artworks that artists had with them already. Klaus wanted to have a travelling exhibition to take with him when he finally went home to Austria. His *Galerie Nowhere* idea was a caravan towed on the back of his bus which would travel from UK to Cuges les Pins in France and then through to his hometown in Graz. During this journey the suitcase artworks would be exhibited randomly wherever Klaus decided to stop.

Due to the variety of social and creative relationships artists had with each other and the public, different currencies of art were produced. Creator of the project, Klaus Mahring, believed that artistic 'activism' was at the heart of the project. The reactionary element he imagined would manifest naturally due to opportunities being contrived for artists to live alternative lifestyles. He believed that providing a place for artists to make work outside the gallery system, in tandem with offering them the chance to meet local people and to share the mutual experience of 'othering', was as close to freedom as one could get in that situation. Also just because Klaus had a vision for the *Nomadic Village* did not mean either that artworks had to adhere to a particular set of standards. For whilst there were no official guidelines about working practises, Klaus chose artists who seemed to subscribe to an unspoken law that collaboration was virtuous and an anti-establishmentarian ethos was attitudinal. In other words the *Nomadic Village* became a platform upon which an eclectic mix of aesthetic hierarchies where able to be explored because the project, despite its restrictions due to funding obligations had no specific expectations. That meant interpretations of space were multifarious. Yet all things produced where viewed with equal measure, whether nomads did nothing other than having conversations during their stay or created things to a very high standard.

That said, in terms of social inclusion, the idea of work made by or with the help of the native community, arguably does not always make 'good' art. Beyond the boundaries of the *Nomadic Village* wall in which equality and democracy resides, a potentially lofty debate about 'high' and 'low' art values is omnipresent. Arguably diverse approaches to creativity meant that a mixed economy of works led to different hierarchical levels in terms of how they could be commoditised. This creative eclecticism would be likely to be questioned by art critics, who may devalue it to the ranks of 'project' art, folk art or 'outsider' art which whilst there are exceptions to the rule, these aesthetic canons are often rated as having a lesser 'value 'in terms of credibility and commercial potential due to them operating on the peripherals of the art establishment, thus not controllable. As the work at the *Nomadic Village* was produced in buses, caravans and ateliers with a focus on expressing lived human experiences within a nomadic construct, the art produced was essentially about nomadic life and interpretation through artistic inter-subjectivities, which because of its ephemeral nature was not that collectable.

The artistic ambition of the project however raises the question as to whether the *Nomadic Village* installation could itself be considered an art form. If the artists have produced an artistic space through its occupation, perhaps a group of artists camping may be considered a curated object. The principle of transforming travelling into art resonates with the Situationalist movement for whom leisure was 'the real revolutionary question' (Sadler, 1998, p. 151). While a paradigmatic example of the 'society of spectacle' that reduces human creativity, the language and practices of tourism are at the very heart of the Situationalist project. In the spirit of Foucault who questioned the idea, 'Could everyone's life become a work of art?' (in D'Andrea, 2007, p. 19), in this instance ignited the debate about whether just being a creative nomad could be in essence, enough to imbue them with status. Quinn suggested that creative conviviality had the potential to transform the currency of people. She felt that the village was like a museum, calling it living art inside a structure that was constantly evolving. She added that even a conversation could be an artwork if defined as such. When other artists were questioned as to whether they thought the *Nomadic Village* was a campsite which accommodated artists or an art installation which looked like a campsite, there was a mixed set of responses, albeit positively inclined towards the former. Fabienne a visual artist from Leon, did think that the place seemed like a peepshow or a visitor attraction to the passers-by. Some of the other artists added that it was not ideal to be thought of as a travelling circus, but conceded that being a public spectacle meant that they could give something back to the indigenous population, who had inadvertently invested in a government funded project.

Political Subversion or a Corporate Event?

There was a fundamental tension at the heart of the *Nomadic Village*. This was a radical project that breaks with the established artistic conventions, in particular the art gallery setting. Yet it benefited from a generous grant from the Arts Council, the endorsement of the local council as well as the onsite 'backseat' management by the public funded organisation ISIS Arts. While the first *Nomadic Village* in Bulgaria in 2009 evolved into an ad-hoc series of comings and goings by artists, friends and locals, the Wolsingham version was not particularly organic in its inception, instead the offspring of a strategic arts development plan. Some scheduling of arrival, departure, mealtimes, project meetings and adherence to contractual obligations was part of the structure with hospitality for the nomads as part of the deal. Provisions such as access to wifi, film screening and presentation facilities were provided as well as the ongoing support from the ISIS whose role onsite was to troubleshoot any infrastructural problems and to deal with public health and safety. So whilst the group maintained an 'alternative' lifestyle as well as a critical stance towards the idea of being part of the system, they conversely enjoyed some of basic 'glamping' luxuries afforded by state sponsorship. Paradoxically whilst *Nomadic Village* offered all of the accoutrements associated with a mobile corporate event, it did so under the banner of facilitating a utopian condition for freedom seeking artists.

Nomadism here is arguably an act of political subversion as well as a lifestyle and a creative tactic. Referring back to the counter cultural movement Situationalism, where a radical view of art has some allegiance, the conservative ideology of the West is perhaps subverted here by the creation of 'situations' where humans interact together as people, not mediated by commodities (Sadler, 1998). Like other radical movements, the *Nomadic Village* aimed to create situations that defy economic rationale. Hannerz (1996) draws the analogy that people who can afford to be open to new experiences and choose to work when they want are comparable with the archetypal writers and painters of Paris between the wars. Cosmopolitans thus, he suggests, can immerse themselves in other cultures by being backstage participants, not front stage passive tourists. The difference at the *Nomadic Village* was that the project embodied a discrete cosmopolitanism, whereby definitions of particular social and political positions were blurred as hosts and guests conversed at different proximities.

The argument that establishing a transnational nomadic community as a powerful critique of hegemonic ideological regimes, has much in common with expressive expatriates in Ibiza and Goa (see D'Andrea, 2007). These hedonistic movements are based on ideas of exile and travel and which actively seek position of marginality from society. Such a connection with other expressive expatriates is not casual. Before the *Nomadic Village* was conceived, Klaus Mahering had been involved with the expressive street party scene of techno nomads, DJ (Sound System) and New Age travellers who would gather by word-of-mouth in often 'unsolicited' spaces. The focus of the *Nomadic Village* residency however was less about hedonism, but more about hybridity. It had evolved as a synergy between neo-nomadic travel as a form of transnational hyper-mobility, creative communal play, yet paradoxically it also had a work ethic akin to professional development associated with any monetarist career motivated pursuit.

The nomadic character of the project was not always in tune with the sedentary logic of art practice and the state. While an opportunity for creativity and subversion, the artists had to comply with regulations and physical boundaries set out by the project. Many of the artists we spoke with noted that due to the way things were set-up, certain freedoms had been compromised. For example Boris, whilst he acknowledged that there had to be some rules, thought that walking around with hard hats on in nature was a bit much. Whilst he was being ironic, one of the other artists, Marek, who was a health and safety officer onsite was observed stepping out of character to perform a comic stereotype of an officious person wearing protective clothing and carrying a clipboard. Boris also agreed that as with any other workplace important regulations had to be adhered to, but remained critical of sanctions on campfires and the heavy-duty risk assessments on tools and equipment due to the health and safety restrictions within the area. Underpinning many of these comments is the wider question of whether nomadic life in the modern world is even possible as a sustainable lifestyle. 'Real nomads are like gypsies', Boris explained, 'people just hate them and they are not allowed to stay anywhere as a matter of their own free will. Because the government want your taxes, people are usually compelled to live somewhere. Nomads cannot exist anymore'.

The *Nomadic Village* was made possible because of the hospitality of Wolsingham Parish Council, who sanctioned the project to be sited on their land. As Sharon Bailey, co-director ISIS Arts pointed out, the local council was very keen to improve tourism within the area so having an International focus was an attractive feature of the plan. The idea of hosting the reality of the *Nomadic Village* in a place not normally used to house event infrastructure raised many questions about the articulation of the local and global as well as the practices of hosting and guesting. The artists were welcomed to settle in the village by the parish council but their hospitality was not unconditional. There was an explicit commitment to give something back to the community in return for being allowed to stay in Wolsingham. As already mentioned, educational projects in local schools and colleges, alongside artists joining in with any social events with the stakeholders, as well as all involved having to embrace the idea of having something to show as a plenary event appeared critical. Artists shared the view that whatever they were doing in Wolsingham, which was by all accounts a fairly conservative rural village, they were adding something potentially 'exotic' to the social fabric of the place. The locals were unaccustomed to being exposed to such cultural diversity inherent in this mobile collective at such close proximity (see Figure 9.2). Whilst

Figure 9.2 Boundaries between locals and nomads

Source: Photo by Sharon Wilson.

it was not tested, their hospitality was perhaps limited in time. If the nomads had decided post-project to overstay their welcome, the tolerance of the residents may have been challenged and the artists once seen as unsolicited visitors may have been dealt with as a nuisance by just being there. The artists in this case however felt that 10 days was the correct timescale for social harmony to be maintained between them and to be able to financially uphold their lives elsewhere before returning to their homelands.

A clear difference was established between the nomadic artists who were assigned the role of cultural ambassadors and other traveling subcultures such as new age travelers or gypsies, generally perceived as freeloaders. Only appropriate types that gave something back to the community were welcomed in a rural landscape. Indeed, the language of art provided respectability to the artistic nomads. The prospect of agreeing to something which had potential caveats, was made more tenable to local officialdom by using the 'ruse' of art. It was a currency for which tourism found valuable and could be used for the education of local communities. It was crucial for the *Nomadic Village* not to be associated with antisocial behaviour. Since the Demense picnic area had never been used officially for camping purposes, the local authorities had to be careful about granting access to projects which by their nature could have been misconstrued. Normally the space in question was as a thoroughfare for local dog walkers and day trippers so for the permission for the arrival of a number of nomadic vans, tents and ateliers, this had the potential to aggravate any unsuspecting local communities who did not attend the consultation meetings or had not seen the publicity about it coming to town. With normative space disrupted, the public could have interpreted nomads in different ways, everything from hostile aliens to intriguing others.

Local engagement with the *Nomadic Village* outside of the pre-booked educational activities was mixed. The majority of people who normally used the picnic area chose to ignore the project and got on with their daily lives as though it was not there. Locals positioned themselves close to the perimeter of the *Nomadic Village* and spent hours enjoying the local river, making barbecues, entertaining their children and drinking alcohol without interacting with artists. Some people did enter the nomadic enclave to find out more and did share in the hospitality of food and drink. Other interactions included local men attempting to form relationships by flirting with nomadic women. Whilst there was no intension on behalf of the artists to be culturally elite, on the event open day only a handful of locals attended. The main audience consisted of primary stakeholders, friends of the artists and the wider artistic community who traveled to Wolsingham for the day to see the final show. As a final caveat, community relations were not without its tensions. In trying to embrace a utopian ideal, artists were unusually trustworthy to others. Laptops and projectors were left unattended and open vans were a constant security risk. Being permissive and perhaps vulnerable was part of the expression that freedom could be enacted. Reciprocity with outsiders it was thought would happen automatically due to an ethical code of peace and goodwill and because it was an open invitation to anyone to find out about the

Nomadic Village; its open nature meant that equipment was stolen and furniture broken, albeit difficult to establish who were the villains. At one stage a meeting was held by Klaus who seemed uncomfortable with leadership on that occasion and found himself managing a situation he thought residents would have policed themselves. In his speech, he was concerned about gatecrashers who partook in the social events too often, which meant that running out of food and the number of times the chemical toilets could be cleaned due to budget restrictions would be compromised. He suggested cutting the number of visitors down regardless of whether they were friends of the artists or local people drifting in. Also realizing potential hazards of the place, he reminded everyone that they had to be vigilant about their possessions as they were not, as he put it, on a dessert island.

Conclusion

In this study, we have looked at an example of a counter cultural artistic practice that brings together travel and art. Unlike traditional nomadic communities, the nomadism of this temporary group is not a survival tactic but another manifestation of expressive individualism. The *Nomadic Village* project funding enabled an artistic elite to be brought together for a working holiday, perhaps paradigmatically not dissimilar to 'ethical' tourists who seek meaningful connections with people and place as part of their travel experience. Whilst conceived as a subversive artistic practice by its creator Klaus, most artists volunteered to be part of the *Nomadic Village* to acquire cultural capital, indeed a volatile currency that is open to multiple interpretations. There is therefore coherence between a volatile system of values that comes and goes and an ephemeral project. For this group of artists travelling was not only about bringing people together but also about a mobile way of life and about creativity inspired by mobility. It was a lens for art produced both as a consequence as well as an expression of transnational movement.

Because of its temporal and spatial fluidity, social formations of this kind are often considered subversive. They are hard to pin down as they are constantly on-the-move. In this instance however it was government sponsorship that made it possible. Politically the nomads were placed in a contradictory position of seeking an alternative lifestyle, whilst complying with an agreed plan of work as well as the rules and regulations associated with occupancy of the campsite. The *Nomadic Village* project focused its attention on the potential of art to disrupt established conventions, although the artists were also happy to participate in established networks. In light of these discussions we therefore conclude with a paradox, for whilst freedom may have been contested in this context, attempts to articulate an aesthetic ideal through the fabrication of a social utopia can perhaps be harmonised between the limited goodwill of travellers in search of the 'goodlife' (Chyutin and Chyutin, 2007, p. 1) and the financial handouts of a 'nanny' state.

References

Chyutin, M. and Chyutin, B. (2007). *Architecture and Utopia: The Israeli Experiment*. Farnham: Ashgate.

D'Andrea, A. (2007). *Global Nomads: Techno and New Age as Transnational Countercultures in Ibiza and Goa*. London: Routledge.

Giddens, A. (1991). *Modernity and Self-Identity: Self and Society in the Late Modern Age*. Cambridge: Polity Press.

Hannerz, U. (1996). *Transnational Connections: Culture, People, Places*. London: Routledge.

Jameson, F. (1991). *Postmodernism or, the Cultural Logic of Late Capitalism*. Durham, NC: Duke University Press.

Kaplan, C. (1996). *Questions of Travel, Postmodern Poetics of Displacement*. Durham, NC: Duke University Press.

Lett, J. (1990). Emics and Etics: Notes on the Epistemology of Anthropology, in T. Headland, K. L. Pike, and M. Harris (eds), *Emics and Etics: The Insider/Outsider Debate* (pp. 127–142). London: Sage.

Levin, H. (1966). *Refractions: Essays in Comparative Literature*. Oxford: Oxford University Press.

Lippard, L. (1999). *On the Beaten Track: Tourism, Art, and Place*. New York: New Press.

Maffesoli, M. (1996). *The Time of the Tribes: The Decline of Individualism in Mass Society*. London: Sage.

Obrador, P. and Carter, S. (2010). Art Politics, Memory: Tactical Tourism and the Route of Anarchism in Barcelona. *Cultural Geographies*, 17(4), 425–531.

Okely, J. (1983). *The Traveller-Gypseis*. Cambridge: Cambridge University Press.

Peters, M., Frehse, J., and Buhalis, D. (2009). The Importance of Lifestyle Entrepreneurship: A Conceptual Study of the Tourism Industry. *PASOS*, 7(2), 393–405.

Pinder, D. (2005). *Visions of the City: Utopianism, Power and Politics in 20th-Century Urbanism*. Edinburgh: Edinburgh University Press.

Rabe, M. (2003). Revisiting 'Insiders' and 'Outsiders' as Social Researchers. *African Sociological Review*, 7(2), 149–161.

Rouse, R. (1988). *Mexican Migration to the United States: Family Relations in the Development of a Transnational Migrant Circuit*. Princeton, NJ: Princeton University Press.

Sadler, S. (1998). *The Situationalist City*. Cambridge: The MIT Press.

Weissleder, W. (1978). *The Nomadic Alternative: Modes and Models of Interaction in the African-Asian Deserts and Steppes*. The Hague: Mouton.

Chapter 10

Tropical Impressions of a Traveller: Caribbean Place and the Retouching of Sensory Experience in the Paintings of Steve Bonner

Marsha Pearce

Introduction

English artist Steve Bonner first visited the Caribbean not as a tourist but rather, as a traveller. The traveller, as writer Gilbert Keith Chesterton (2006) tells us, 'sees what he sees' in contrast to the tourist or 'tripper', to use Chesterton's concept here, who, through a lens of expectation, 'sees what he has come to see'. Bonner had no plans to go see the Caribbean, but in 1995 he took the opportunity to travel to the islands, specifically to the twin-island Republic of Trinidad and Tobago, in support of friends who were on a cricket tour. On his return trip to England, the plane stopped in Barbados. What he saw of that island, through the airplane window, was unexpected; it left an impression on him, it touched him. According to Bonner:

> It was on that very first trip – I was lucky enough to have a port side window seat – and as the plane took off and banked, as it does, around St Philip [a parish on the south east coast of Barbados], I looked down on all the houses, all the plots, and in just a few seconds I knew – I knew that somehow I had a destiny that involved this little island. (personal communication, September 15, 2012)

This chapter pivots on notions of seeing and touching as it investigates what Bonner saw and continues to see and paint on his return trips to Barbados, trips made in order to produce art that is primarily targeted at vacationers looking for visual reminders of their island sojourns (see Figure 10.1). The chapter insists that his paintings of Barbadian landscapes, while not in keeping with the technical style of nineteenth-century Impressionism, are phenomenological impressions, which let us see what he sees. It argues that his representations of Barbados constitute a retouching of his sensory experiences of a Caribbean place. In other words, retouching is a creative effort, through the application of paint on canvas, to make contact again or to reconnect with, to 're-touch', sensations or impressions of place.

Figure 10.1 *Surf and Spray*, Steve Bonner, oil on canvas
Source: © Steve Bonner, reproduced with permission.

Bonner speaks about this need for retouching and, in doing so he makes a distinction between the visual modalities of photography and painting and hints at the implications of this distinction for travelling and image-making. On his website, Bonner shares:

> Have you ever taken a photograph of a mountain, and then been disappointed with the result? It all seems smaller than you remember, or the light and colour isn't as vibrant. Your senses register beauty, awe, majesty, magnificence – all of which, if not actually lost in a photograph, are certainly diminished – my job, the job of the painter, is to put that back. (2012)

He intimates a difference between the point-and-shoot-traveller who relies on the eye of the camera to see and capture place and the painter-traveller who relies on his own eyes. He suggests that in acts of pointing and shooting, there is a loss of the experiences of place in the photographic image that is produced, that is, a sensation of place is 'mis-placed' somewhere between the camera's eye or lens and the camera's memory in the technological sense. However, the act of painting, in which the painter-traveller manoeuvres between the sensation of sight (human eye and brain) and canvas, remedies misplacement through a retouching of the painter-traveller's own sensory impressions or memories.

In the context of this chapter then, retouching is analysed as a form of remembering, sense recovery or 'putting back' and a forestalling of experiential loss. The term is taken beyond the confines of a present, prevailing equation with 'Photoshopping' or the digital altering and manipulating of images in post-production (that is retouching after a photograph has been taken or produced) to its role within the very act of primary production: the construction of painted impressions of place. This investigation is based on both interviews and visual data. It draws on information from personal communication with the artist and incorporates his artwork as points of analysis. The chapter moves through a series of conceptual steps, from considerations of reality and connotations of retouching to the impressionist art movement, providing the contextual foundations upon which to review Steve Bonner's artworks and art-making process. An examination of some of Bonner's paintings, accompanied by insights extracted from interviews with the artist, constitutes the latter part of this chapter.

Seeing Realities of Place

For Steve Bonner, his artist's eye engages with places in Barbados by deploying a particular kind of optics: he 'sees the reality and not the actuality' of a place (personal communication, September 15, 2012). His words call for explication here. In describing his visual experience of Barbados, Bonner takes up a very specific philosophical perspective, which carries with it, a particular orientation.

In the philosophy of science for example, different philosophical perspectives or paradigms are usually understood as 'way[s] of looking at the world' (Mertens, 2010, p. 7, letter in brackets added). Paradigms comprise of certain philosophical positions or notions about the world around us, which are characterised by responses to a number of questions including: What is the nature of the relationship between the knower and the world (or the knowable)? What is the nature of that which the knower comes to know? (Guba, 1990; Guba and Lincoln, 2005). A rationalist or positivist philosophical position holds that such a relationship should be objective or impartial and that what the experiencer comes to know about the world should be 'factual' or 'actual'. An interpretivist or constructivist position takes the view that the relationship between the knower and the world is relative and that what the knower knows or 'realises' is subjective. In their discussion about how we gain knowledge about the world Marton and Booth maintain: 'People … are affected by what affects *them* … we do experience the world differently …'. (1997, p. 13, emphasis in original). Hall (1997) echoes this view. While recognising that we are able to have a shared culture of meanings because we can see the world in roughly similar ways, he says that 'In fact, each of us probably does understand and interpret the world in a unique and individual way' (1997, p. 18).

Therefore, if these philosophical perspectives are taken into account when Bonner says that he sees the reality and not the actuality of a place, he speaks in constructivist terms. He articulates an experiential relationship with Barbados or a visual engagement with the island place that is not objective. He takes the position that we interpret the world around us, producing our 'realisations' or perceptions of it. His sensory experience therefore does not yield the actuality or fact of place but rather his perceived reality. Carey observes that 'reality is not given … it is for us a vast production … something humanly produced' (2009, p. 20–21). Reality, Carey argues, is constructed in language. Reality is 'brought into existence' through languages or 'symbol systems' like art (2009, p. 20 and 24). It is his realisation of Barbados, that is, the reality he sees, that Bonner attempts to bring into existence in his art.

Bonner organises paint on canvas so that it represents those features of the sea, rocks, trees and other details of island place as he has seen them. Each painting is a subjective construction and a marker of his subjectivity. In her discussions of visuals in tourism research Rakić (2012) notes that images are infused with subjective, human viewpoints in their making. She writes: 'images of any type … are inevitably … made with a specific purpose by an inevitably subjective historically, spatially, culturally and socially situated individual …' (2012, p. 25). While not denying that Bonner's artworks are produced for sale in a tourist market, he has his own 'specific purpose' in his image-making practice. The artist creates images with a key intent to remember his subjective experiential moments, to access his memory, to return to, reconnect with or to retouch his realities of Barbados.

Practices of Retouching

The practice of retouching can be traced to notions of 'remembering' and 'return'. Retouching has long been a process of restoration in the arts. Paintings that have become discoloured or dull over time due to light exposure, extremely moist conditions, dust and other contaminants are retouched. The act of retouching in this sense is a remembering of and return to the painting's original state as the restorer works to re-establish colour and detail where there are losses or lacunae in the paint layer. In our contemporary digital context, retouching has become a more aggressively pursued act with the ubiquity of digital cameras and image editing software. Yet, the meaning of retouching has taken on another inflection.

Images are made with the push of the camera's shutter-release button and they are instantly saved on the camera's memory card for later access. Once retrieved, images are often digitally retouched, in other words, they are altered, edited or enhanced. Using such software as Photoshop, Pixelmator or PaintShop Pro, adjustments can be made to brightness, contrast, sharpness, hue, saturation, shadows and highlights among other pictorial modifications. Elements may also be added to or removed from an image. The result of this post-production process is a new image, one that is different from that which was first stored in the camera's memory. Retouching, in this sense, is more of a creative decoupling from rather than a reconnection with, or restoration of, what the camera's eye saw and captured in its memory bank. What is stored in memory, more often than not, becomes a point of departure rather than a point of return.

While addressing the same subject matter, the image that is stored in the camera's memory offers one reality and the retouched digital image gives another. The relationships between the camera, acts of retouching and the issue of reality are significant to tourism practice, wherein Sontag (2002, p. 177) notes a 'dependence on the camera as the device that makes real what one is experiencing'. The tourist relies on the camera to realise experiences or, in other words, to produce perceived realities from experiences, however landscape photographer Briot insists that what the camera's eye realises and what the human eye realises are usually vastly different:

> The fact that the camera stands between the photographer (us) and our subject means that a number of changes will be introduced … nearly all lenses introduce some form of distortion … blurriness, chromatic aberration, vignetting and the like … . Film or digital sensors also introduce a number of changes such as film grain, noise … pixelization, color shifts, color casts and so on … each film or digital sensor has a specific contrast range and color gamut. If this contrast range or color gamut are exceeded by either the brightness range of the scene or the intensity of the colors in the scene, clipping will occur and the photograph will suffer from featureless shadows and/or blown highlights. (2012a)

The camera mechanism imposes 'changes'. Therefore, if we are not professional photographers armed with the technical expertise to manipulate camera elements in an effort to produce images that will match our visual experiences of places more closely, the camera, as Briot adds, can be 'damaging to the reality we see' (2012b). His use of the word 'damage' can be interpreted as loss, a loss of the dimension of our own sensory experiences and loss, or mis-placement, of our experiences of place. Travel blogger Forshner Frechette (2011) alludes to this loss. In her review of a travel photography book that offers tips for capturing travel experiences, she shares what she wants to get from her photographs of places she has visited:

> I'm someone who simply wants to capture my travel adventures in a truly vivid way. I'm not out to become a professional travel photographer, but I always want to walk away from a trip with more than a slew of one dimensional shots of landmarks. I want my experiences captured. (Forshner Frechette, 2011)

Her reference to the potential for leaving a trip with one-dimensional shots of place points to the camera's ability to lose or damage sensory experiences and rule reality in the hands of the amateur photographer or point-and-shoot tourist. How then, might such a loss be restored? How might this damage be repaired? How might the tourist retouch his or her own realities of place?

Trend (2011) broadens this thesis of loss in his piece, 'Travel Photography: Escaping the Tyranny of the Camera'. Trend argues that the camera can make tourists less observant. He contends that the camera can interrupt our memories and compromise our experiences of place. Trend tells us:

> It's easy to think of photographs as an aide-mémoire, but do you remember taking each of them? I certainly don't. Many are capturing moments I have forgotten. I recognise the scene and the people, of course, but I'm not sure I really remember experiencing the moment itself. I think photography stops you looking properly, and interrupts the emotional experience of seeing new and exciting things. The syndrome has been intensified by digital cameras. Now you don't even look through a lens at your subject; you look at an electronic display screen. (2011)

Trend proposes a loss of seeing and experiencing places with our own anatomical mechanisms, that is, our own sensory devices. In his call for us to disentangle ourselves from the camera's power, he uses an exhibition of travel paintings to make his point about the capacity to see and remember more through the practice of painting. Trend refers to a watercolour exhibit, which was held at the Tate Britain museum from February to August 2011. The travel and topography section of the display featured eighteenth-century artists who, from their travels, produced paintings from what they saw, artworks that responded to the spirit of place. Trend maintains that in lieu of photography these artists were 'intensifying the moment and … their memories' (2011). Such paintings could forestall a loss of sensory experiences or thwart a damaging of the realities perceived by the travelling artist.

It is with this understanding of a potential for the fading of the intensity of moments, the forgetting of experiences and the likelihood of not seeing with our own eyes that we can deem a possibility of loss or damage, that we can reconsider a notion of retouching as a kind of re-association with restoration. Steve Bonner can be seen as a restorer. His artistic practice is an attempt to restore his memories of place in each painting. Retouching is not an act of post-production for him, as is the case in digital imaging. Instead, his work is characterised by a retouching in the primary production of a work of art. What he paints does not amount to something different from what he has stored in his memory. His paintings are a retouching, a remembering or a return to a reality he has perceived. 'When I paint, I aim to paint that very first impression', says Bonner (personal communication, September 15, 2012).

Bonner's Art and Nineteenth-century Impressionism

Steve Bonner seeks to paint the effects that place has on him, the impressions a place makes on him. This aim is in keeping with the tenets of a group of artists, which would later be called the Impressionists, who first exhibited together in France in 1874. They emerged in the wake of the French Revolution, in a creative climate typified by an austere form of painting with strong linear qualities and sombre colours. Impressionists would respond subversively in such a context, liberating themselves from an attention to line and using vibrant colours to capture the impact or effect of people and place on their senses. Thomson explains what the Impressionists were trying to accomplish:

> ... what the artists were seeking to translate was not so much an externally verifiable standard of truth to appearance as their own ... response to an aspect of nature at a given moment. The term the artists themselves used for this was 'sensation'. In their parlance, sensation came to mean more than a physiological response to visual stimuli; it became a highly personal response to a particular effect. (2000, p. 192)

Impressionism is tied to the sensation and effect of the moment. Graña observes that Impressionism 'assumes a world in which moments can exist as total units of experience: where self-feeling ... has a new swiftness and, within that, a new, flickering poignancy' (in Herbert, 2007, p. 24). Impressionism, according to House, refers 'to the impression made on the viewer by the experience of nature's most transitory effects and then, by extension, to paintings that recorded such impressions' (2007, p. 87).

It is in this vein that Steve Bonner calls himself an impressionist painter. His work however is not in keeping with what is considered to be an impressionist execution of a painting. His paintings do not bear such attributes as indistinct contours and visible brushwork made with alternately long and short strokes of paint (technical qualities that are evident in such works as Claude Monet's 1875 painting entitled *Les Bateaux rouges, Argenteuil* and Camille Pissarro 's 1889 piece *The Gleaners*). Bonner elides such stylistic features in his representation of place, preferring to paint with his own methodology while holding fast to a want to communicate what he has experienced. In talking about his approach to his art, Bonner shares what brought him to a point in his creative practice where he understood what he was trying to achieve in his impressionist paintings of place:

> I think for me the penny dropped when I was looking at [J.M.W.] Turner's paintings of the North Yorkshire moors. These are views one sees constantly in the UK, in magazine ads, or on TV programs, and when I saw the paintings I thought to myself: 'They don't look like that. The cliffs in the paintings are far too grand!' But they're not, you see. When you actually visit Yorkshire, when you stand beneath those sodding great cliffs, you can't help but

> go 'Wow!' and it's the wow that Turner painted! In my Barbados paintings, it is – I hope – that 'wow!' I am putting back. (personal communication, September 15, 2012, text in brackets added)

Putting Back Experiences of Barbados

After Steve Bonner's first trip to the Caribbean in 1995 for a cricket tour (a visit during which as mentioned earlier, he spent his time on the islands of Trinidad and Tobago), he returned to England with plans to create a body of work: reproductions of Tobago for the tourist market. He returned to the islands the following year where, as he describes it, his work was well received. On his departure, the plane again touched down in Barbados. Bonner shares, 'still my only experience of the island was the airport terminal and that same curiously persistent nagging feeling as we took off' (personal communication, September 15, 2012). Bonner would soon make a more tangible connection with Barbados, through the advice of a friend, a merchant banker, who offered to help him with his efforts to paint and sell images of the Caribbean. According to Bonner:

> [My friend] explained that Trinidad and Tobago, while a major player commercially within the Caribbean, has a comparative small tourist industry and so, from our point of view, Barbados was a much better bet. Thus, 1997 found my wife and I, camera and paint box in hand, walking out of [Barbados's] Grantley Adams International Airport into the Bajan sunshine. (personal communication, September 15, 2012, text in brackets added)

Bonner has since returned to Barbados on a number of occasions over the years in order to paint the place. His returns to the island are part of his painting process, for he makes the trips back to the Caribbean to retouch its heat. Many times, Bonner starts a painting in England, where he has more time to accommodate long waits for oil paint to dry on the canvas before he can apply more pigment to the surface. In England he can also support himself financially while making works of art but he prefers to finish his paintings in the Caribbean. He says:

> There is a strange phenomenon with the heat in the painting … . If I paint in the UK … the painting looks cool when hung in the gallery. Now I know that at home … there is more blue and less red in the visible spectrum, an effect I presume, of the filtering effect of atmosphere with the sun at a more oblique angle. But if I use the self same palette, the self same tube of paint even, why does a painting I do in England still seem cooler than one I paint in the Caribbean? Logic would suggest that I strengthen the colours to compensate, which I'm sure I do. [It is] all very curious. (personal communication, September 15, 2012, text in brackets added)

Bonner expresses an inability to recall and record on canvas a phenomenological Caribbean temperature, with the distance of the Atlantic Ocean between him and the islands. He must travel back to remember or retouch the warmth of Barbados, to reconnect and paint the reds and crimsons of his sensory experience of island place. For Bonner, completing paintings in the Caribbean 'makes the work feel more as if it belongs' to the tropical place (personal communication, September 18, 2012).

More than a connection between painting and place, this idea of belonging also extends to Bonner's own affiliation – as an artist-traveller – with Barbados. He finds that he can connect, again and again, to place through the process of painting, which allows him to reactivate his personal experiences with Barbados's landscape and have deep and intimate attachments to the island. Bonner speaks about the transience and fleeting connection a tourist can have with a place and in contrast, he foregrounds the enduring links he makes with place and the depth of vision he employs through art-making:

> A tourist will fly in, make observations, maybe a few judgments, enjoy the sun and fly home … . I have always been aware of the need to get 'under the skin' of a country, wherever I go. Being a tourist is a foul experience, I hate it, but my painting enables me to 'belong' to a degree … . The very act of painting, of moving the brush over the canvas, creates an intimacy, a 'oneness' with the subject matter. (personal communication, September 18, 2012)

It is this 'oneness' with the place, a retouching of his sensory impressions of Barbados, that manifests on canvas. His is an effort to create a unity between what he paints and what he has experienced on his trips.

For Bonner, getting 'under the skin' of Barbados involves personal realisations of what he calls the awe and majesty or the 'wow' of Barbados, realisations that attempt to get beneath the epidermis of conventional pictorial tropes of island sun, sea and sand, while interestingly, seeking to appeal to a tourist market that typically feeds on visual stereotypes of tropical places. It is, perhaps, Bonner's capacity to make both tourists and the people of Barbados confront what has been unrealised in their own experiences of the island, his creative ability to expand their consciousness and see the place anew, that has allowed him to sell as many of his oil paintings to island locals as he has sold to tourists. Each time he puts his brush to the canvas, he attempts to retouch and put back his realisations or realities of Barbados.

In his piece entitled *Chalky Mount* (see Figure 10.2), Bonner aims to remember his experience of a rugged range of hills located in the Saint Andrew parish, in the eastern area of Barbados. In retouching this island location he attempts an act of restoration, that is, a return to and preservation of Chalky Mount as he has realised or perceived it. He realises its grand and imposing scale for instance, in his sensory engagement with the site.

Figure 10.2 ***Chalky Mount*, Steve Bonner, oil on canvas**
Source: © Steve Bonner, reproduced with permission.

He aims to render his reality of Chalky Mount with as little damage as possible. Yet, to preserve that reality, to sustain his experiences of the location, he must edit and manipulate Chalky Mount in a manner that makes the notion of retouching resonate simultaneously with hints of early and contemporary meanings of the word. Bonner explains, while making reference to issues of loss and putting back:

> This picture is an example of manipulation to put back the 'awe and majesty,' which a holiday snap removes. The 'mass' of the mount has been increased considerably without actually stretching it. It probably isn't the best composition ever, which is deliberate, as extending the range to the either side would diminish its impact. The house, poles and cables in the foreground have been removed and the cows added to give a sense of scale. Have Ospreys ever soared above Chalky Mount? I have no idea – but they do in my world. (personal communication, September 19, 2012)

That which exists in actuality: poles and cables in the foreground, are edited out in favour of what has made an impression on him or of what he is more conscious. His 'manipulation' does not give rise to a different reality for him. Rather, it reinforces what has been most forcefully recorded in his perceptual reality or 'world', as he puts it. While an osprey may not have had material presence in his experience of Chalky Mount, the bird comes into existence for Bonner as a signifier that matches the intensity of the moment. In his reality, the powerful soaring flight of an osprey, with its noticeably big wings, magnifies the rocky mound, which Bonner senses as powerfully colossal. As a bird with a particular species that can be found worldwide, the osprey is not out of place in the painting. It is a plausible representation of the tropical place. The almost imperceptible cows in the piece communicate the relative size of the mount and the magnitude of the artist's experience.

In *Low Tide Cattlewash* (see Figure 10.3), Bonner retouches his experiences of the energy and currents of a beach location on the eastern coast of Barbados. Cattlewash beach is one of the longest stretches of beach in Barbados. It is also less populated by sea bathers because of the extremely strong Atlantic waves that pound the shore and beat upon the row of rocks that characterise the beach.

Bonner engages with Cattlewash on a day when the tide is low and the sea is considerably calmer but in his painting he puts back what he has sensed and realised on other occasions: ocean as far as his eye can see, a mass of foam breaking on a rock and the overwhelming impression of rows of rocks dominating his field of vision. He tells of the devices he has deployed to remember and reconnect with his experiences of the vitality and dynamism of the beach:

> The first thing to attend to was the sky … . To emphasise the ocean on the left of the painting, I chased the clouds back across to the land … . Secondly, there are two rocks some yards further back than the foreground chain [of rocks]. From the exact viewpoint of the picture those two rocks are only partly visible – I've shifted them into clear view … . When one comes upon that scene your sight is FULL of rocks – they continue to march up the beach. You just don't stop to consider what goes where. In the painting I have to try to recapture that. Also, there was very little surf that day and yet the outermost rock [in the painting] has surf breaking over it – another device. I have to remind the

Figure 10.3 ***Low Tide Cattlewash*, Steve Bonner, oil on canvas**

Source: © Steve Bonner, reproduced with permission.

> viewer how those rocks have become so weathered. So despite the calm day, and the smooth low tide, I still have the surf in the painting. I have also increased the presence of the rocks in the foreground and slightly closed them together (personal communication, September 29, 2012)

Bonner revises the image in order to restore and 'recapture' his sensory experiences. Along with shifting items, adding features and increasing the size of picture components, Bonner acknowledges that a key factor in his practice of retouching is the choice he makes regarding the size of his canvas:

> One element, which has helped more than any other with my Barbadian landscapes – crucial in fact – is the format of the canvas .... *Low Tide Cattlewash* was painted on a [canvas] 40 x 20 inches .... I work quite hard at recapturing the visual impact of the scene I'm looking at, but that only works because the canvas itself allows me to do so. Imagine *Low Tide Cattlewash* on say, 30 × 40 inches. Already, I will have physically ten inches more sky or sea. Whatever I do to my ribbon of rocks [at that canvas size], unless I almost cartoon them, they can't but be diminished in the viewer's eye. (personal communication, September 29, 2012)

The artist also makes use of a wide format canvas, 40 × 20 inches, in his piece entitled *Swallow's Playground, North Point* (see Figure 10.4). It is a painting that emerges from Bonner's experience of a tourist attraction in Barbados known as Animal Flower Cave, where sea anemones can be found in the pools of the cave. Bonner shares that by visiting the site, he was 'playing tourist' that day, in other words, he performed the role of one who visits places of interest that are mapped out in tourism brochures and magazines.

Figure 10.4 ***Swallow's Playground*, North Point, Steve Bonner, oil on canvas**

Source: © Steve Bonner, reproduced with permission.

During his visit, he transgressed the customary tour and against his guide's advice, he stepped out of the cave onto a coral shelf where the waves nearly washed him away. There he got a high-angle view of the cliffs and caves of North Point. He experienced a magnificence, which he attempted to capture with a camera. He tells of the damage the camera did to his reality of the moment:

> The snaps I took were a disappointment. Everything was lessened. But in my painting, I hope, it's restored to its magnificence. Looking at the photos I had no idea of scale, no idea that these caves are so huge they could swallow a house – a big house – so that is what I had to put back into my painting. (personal communication, September 29, 2012)

Though he was standing above the view, he chose to paint it at a level that allows the caves to tower above the viewer. Again, as with other paintings, he gives a sense of the grand scale of the caves by including two tiny figures in the top right. In his construction of the piece he attempts to retouch or remember his sensory experiences:

> I pushed the caves back and pulled the cliff forward – that's how I remembered it. My photos told me that the next bay was quite close, as was the next, and the next but that wasn't how I remembered it, so I slashed the perspective back. All I remember, as I stood on that rock shelf up to my ankles in swirling sea water … was spray drenching me, and right before me, massive, gleaming, coral cliffs and vast caverns that filled my vision – that's what I wanted to paint. (personal communication, September 29, 2012)

Conclusion

Using the theoretical frame of subjective representations of experience, this chapter has insisted that Steve Bonner's paintings of Barbados constitute a retouching of a phenomenological Caribbean place. In other words, his art is a means of reconnecting with his visual sensory experiences of the island and representing those subjective tropical impressions. Steve Bonner's art makes his memory of the place tangible. He executes a creative method built on a foundation of retouching that takes him back to moments and resuscitates his own personal realities of the island place.

A look at Bonner's creative practice exposes the important role art can play in expressing our experiential relationship to places, in helping us 'return' to the places we have visited, sustaining our memories and perceived realities of those places. Examining this process also reveals how art can be a powerful motivating force for travel as demonstrated in his need to make trips from the UK to the Caribbean in order to retouch and represent a tropical heat in his paintings. Yet, Bonner's work carries particular significance because it highlights a complex nexus of art, tourism and travel. The artworks, which are constructed through various devices and manipulations employed by this painter-traveller, operate not only as vehicles for personal statements of remembrance but they also become caught up in a matrix of visual representations of the Caribbean (including postcards and travel magazine photographs): a dominant visual matrix with a long history wherein the Caribbean has become what Sheller (2003, p. 65) calls 'an invented landscape' for tourist consumption.

References

Bonner, S. (2012). About the artist – Steve Bonner: A Biography. Retrieved 5 June 2012, from <http://www.stevebonner.co.uk/steve-bonner_biography.html>.

Briot, A. (2012a). Art and Science. Retrieved 23 September 2012, from <http://www.luminous-landscape.com/columns/Art-Science.shtml>.

Briot, A. (2012b). The Eye and the Camera. Retrieved 23 September 2012, from <http://www.luminous-landscape.com/columns/eye-camera.shtml>.

Carey, J. (2009). *Communication as Culture*. New York: Routledge.

Chesterton, G. K. (2006). *The Autobiography of G. K. Chesterton*. San Francisco: Ignatius Press.

Forshner Frechette, S. (2011). Capturing the Journey – Travel Photography Ebook Review. Retrieved 18 November 2011, from <http://transienttravels.com/2011/10/19/capturing-the-journey-travel-photography-ebook-review/>.

Guba, E. G. (1990). The Alternative Paradigm Dialog, in E. C. Guba (ed.), *The Paradigm Dialog* (pp. 17–27). London: Sage.

Guba, E. G. and Lincoln, Y. S. (2005). Paradigmatic Controversies, Contradictions and Emerging Confluences, in N. K. Denzin and Y. S. Lincoln (eds), *Sage Handbook of Qualitative Research* (pp. 191–215). London: Sage.

Hall, S. (1997). The Work of Representation, in S. Hall (ed.), *Representation: Cultural Representations and Signifying Practices* (pp. 13–64). London: Sage.

Herbert, R. (2007). Impressionism, Originality and Laissez-Faire, in M. Tomkins Lewis (ed.), *Critical Readings in Impressionism and Post-Impressionism* (pp. 23–29). Berkeley: University of California Press.

House, J. (2007). Framing the Landscape, in M. Tomkins Lewis (ed.), *Critical Readings in Impressionism and Post-Impressionism* (pp. 77–99). Berkeley: University of California Press.

Marton, F. and Booth, S. (1997). *Learning and Awareness*. Mahwah: Lawrence Erlbaum Associates.

Mertens, D. (2010). *Research and Evaluation in Education and Psychology*. London: Sage.

Rakić, T. (2012). Philosophies of the Visual (Method) in Tourism Research, in T. Rakić and D. Chambers (eds), *An Introduction to Visual Research Methods in Tourism* (pp. 17–32). London: Routledge.

Sheller, M. (2003). *Consuming the Caribbean: From Arawaks to Zombies*. London: Routledge.

Sontag, S. (2002). On Photography, in D. Crowley and P. Heyer (eds), *Communication in History: Technology, Culture and Society* (pp. 174–178). Boston: Allyn & Bacon.

Thomson, B. (2000). *Impressionism: Origins, Practice, Reception*. London: Thames and Hudson.

Trend, N. (2011). Travel Photography: Escaping the Tyranny of the Camera. Retrieved 23 September 2012, <from http://www.telegraph.co.uk/travel/travel-advice/8575416/Travel-photography-escaping-the-tyranny-of-the-camera.html>.

Chapter 11
Martin Parr: A Traveller–Critic and a Professional Post-tourist in a *Small World*

Theopisti Stylianou-Lambert and Elena Stylianou

Introduction

When thinking of the world, one usually thinks of it as something large, full of unknown, distant, and at times exotic places that motivate people to travel. Encouraged by the desire to see and experience new destinations, have adventures, relax away from everyday life and, ideally, away from other tourists, people tend to sustain the notion that the world is indeed large. Yet, the celebrated British photographer Martin Parr who often travels to observe and photograph tourists sees the world differently. In his book *Small World*, which includes 74 colour photographs of tourists at different locations, Martin Parr records various elements of the tourism industry. The title of his book seems to be a reference to the notion of globalised travel and to those technologies that shrink distances and make accessible all the things and places that were previously remote. At the same time, it also refers to the homogenisation of behaviour among tourists who, regardless of nationality, gender or past experiences, replicate and perpetuate certain patterns of behaviour when travelling. This type of accessibility, shrinking of geographical boundaries, and similarity in behaviour noticed by the photographer, not only make the world seem rather small and superficial, but, as we will see, also recall modernist approaches to tourism.

The photography book is considered by Parr to be the ideal vehicle to publicise his work because 'it's printed, it's reproducible and it travels well' (Lane, 2006, p. 15). The book *Small World* offers brief visual case studies that reveal Parr's understanding of tourism in relation to global culture. First published in 1995 and then re-published with additional photographs in 2007, it is the first of a series of similar photo books and was followed by *Tutta Roma* (2006), *Mexico* (2007a), *Machu Picchu* (2010) and *Japan* (2011). In all of these books, Parr takes pictures of tourists at famous locations, but purposefully shifts the emphasis from the locations and landmarks to the tourists and the tourism industry. *Small World* is not limited to a specific location and contains most of the thematic and visual elements that appear in many of Parr's subsequent books, so it is the ideal starting point for analysing his work and views about tourism.

Parr considers tourism as the biggest industry in the world and explains that what interests him is 'the great conundrum, the contradiction between the mythology of these places and the reality' (in Romig, 2010). In his introduction of Parr's website, Thomas Weski states: 'Parr's term for the overwhelming power of published images is "propaganda". He counters this propaganda with his own chosen weapons: criticism, seduction and humour' (Weski, 2013). His images, often dramatically different from the happy, excited or relaxed lone tourist encountered in promotional material, effectively point out what is omitted from the untroubled world of commercial imagery, bringing to the surface both the realities and the politics of tourism. While people usually travel to experience a foreign destination, to relax, explore, enjoy, socialise and be surprised, Parr's tourists often seem to be lost, bored, anxious or tired. They are captured buried behind maps or camera lenses and engaged in group activities.

The ways we read and interpret the work of an artist are influenced by particular theoretical framings. This chapter demonstrates how modern and postmodern theories have an effect on the way the work of Martin Parr is conceptualised regardless of the artist's intentions. Until recently, Parr's work has been framed within modern theories of tourism, those that see the tourist experience as mass, trivial and superficial. Indeed, as we will discuss in this chapter, Martin Parr's photography often highlights a passive tourist, an inauthentic tourist stage, an artificial, problematic tourist-host relationship and the strange conventions created to facilitate tourism. Perhaps for this reason, various theorists have used Parr's work as a way to reinforce the normative critique of modern mass tourism, which began with Boorstin (1987 [1961]) in the mid 60s. Nonetheless, Parr's role as an artist documenting tourists and tourism practices can also be seen and discussed differently when considering postmodern tourist theories, and by doing so, it is hoped that a new understanding of his photographic practices can be introduced. Influenced by postmodern theories, we are suggesting a new reading of Parr's *Small World.*

The Tourist: Modern vs. Postmodern Theories

Since the 1970s the tourism field has been dominated by two competing approaches to the tourist experience (Uriely, 2005). One approach was exemplified by Boorstin's work (1987 [1961]), which saw tourism as 'a symptom of modern decadence' (Uriely, 2005, p. 982) and the tourist experience as superficial. On the other hand, the opposing approach represented by MacCannell (1999 [1973]) saw the tourist experience 'as a meaningful modern ritual which involves a quest for the authentic' (Uriely, 2005, p. 982). MacCannell's work opened the door to postmodern theories, which were non-dualistic, anti-hierarchical and placed the emphasis on tourists and their multisensory experiences, instead of their behaviour or the factors that structure their gaze (Urry, 1990).

The 1990s saw the emergence of a new approach. The 'performative turn' (Haldrup and Larsen, 2010, p. 3), as it was later named, saw tourists as performers who are actively involved in the construction of visualities and identities and are thus granted the status of cultural producers in addition to that of cultural consumers (Bærenholdt, Haldrup, Larsen, and Urry, 2007; Crouch and Lübbren, 2003; Edensor, 2000; Larsen, 2005; Robinson and Picard, 2009; Scarles, 2009; Stylianou-Lambert, 2012). Terminology borrowed from theatre, such as 'performance', 'actor', 'stage' and 'choreography' was deemed appropriate to describe the newly discovered activity of the tourist (Edensor, 2000). This emphasis on performativity, a purely postmodern approach, seems to empower tourists by focusing on social relationships, the production of myths and narratives, and tourists' photographic processes.

It is within this shifting paradigm from modernity to postmodernity that Martin Parr's practices become increasingly intriguing and relevant. In what follows, Parr's book *Small World* will be analysed within the context of both modernist and postmodernist discourse. Our analysis relies on the principles of semiotic analysis of images contained in *Small World* along with a consideration of various texts which mention Parr's work, including his own commentary. As we will see, depending on whether we adopt modernist or postmodernist views of tourism, the artist can be envisioned as a 'traveller–critic' or a 'professional post-tourist', respectively. However, the one view does not necessarily contradict the other. On the contrary, new readings should be seen as layers of complexity added to the artist's work.

Modernist View of Parr's Tourists

The Mass, Global Tourist

In his 1961 book *The Image: A guide to Pseudo-Events in America*, Boorstin (1987 [1961]) laments the disappearance of the 'true', active, risk-taking, adventurous traveller and the emergence of the mass, passive, pleasure-seeking tourist. The nineteenth-century Grand Tour leisure traveller had the privilege and capability to observe, take notes, sketch, spend time noticing local manners, and collect souvenirs. The tourist later replaced this traveller. As Fussell describes in his book *Abroad* (1980), tourists only travel in 'conspicuous crowds' and at certain seasons, and they are only interested in the 'security of pure clichés' rather than taking the risks of exploring the unknown (in Kaplan, 2005, p. 53–54).

Parr seems to visualise both Boorstin's and Fussell's modern profile of the tourist by emphasising, and subsequently also criticising, the global and mass character of tourism in many of his photographs. In one photograph in *Small World*, for instance, Parr captures a large group of tourists, mostly Asian, lined up for a gondola ride in Venice (Parr, 2007b, p. 18). This sight, in sharp contrast with the notion of a romantic gondola ride for two, captures the stereotype of the tourist with the camera, hat, flat shoes and eagerness to wait in line in order to consume a top-ten must-do experience, as described in travel guide books. Parr's tourism is presented as a collective event, a social, international ritual (Osborne, 2000).

Similarly, in another instance, Parr photographs the cramped passengers of a plane simultaneously raising both hands in the air as on cue (2007b, p. 95). Some look excited, some forced, but the uniformity of what seems to be a choreographed gesture can be seen as both humorous and sarcastic. In these images, the photographer is not just recording the tourist behaviour, he is also criticising it as often being silly and driven by mere desire to have fun, rather than being the result of an honest interest in learning, experiencing or discovering new things at the visited location. The depiction of crowds in his photographs connotes just that: the tourist is not an interested, thinking individual, but simply part of a crowd and as such he/she only follows directions. Parr's visual conceptualisation of the tourist is not far from general academic orthodoxy, which often views the tourist as a 'superficial, materialistic, ignorant, insensitive, rude, and noisy hate-figure of polite society and academic discourse' (van de Ven, 2011, p. 142), a subsequent result of the relation established between photography and the colonial gaze (van de Ven, 2011).

The Tourist's Gaze

Nothing is more telling of a tourist than the camera usually found around his/her neck. Thirteen out of the 74 photographs in *Small World* (including the cover) feature people posing or taking photographs while many more include a camera in the picture. For example, in one photograph we see a man in Kalkan, Turkey, riding a horse guided by a local who walks gingerly next to the horse (Parr, 2007, p. 2). Both tourist and host look straight ahead, only the tourist filters his gaze through the lens of what seems to be an uncomfortably large video camera. This image humorously poses questions regarding the kind of experience that a tourist can have with a camera. One might claim that the tourist riding the horse sacrifices the actual experience for the sake of accumulating images for future consumption. According to Urry and Larsen:

> The person who only lets the sense of sight have free rein is ridiculed. Such sightseers, especially with a camera draped around their neck, are conventionally taken to be superficial in their appreciation of environments, peoples and places. Martin Parr's photographic collection *Small World* reveals and exposes such a destination of the (normally male) camera-wearing tourist (2011, p. 18).

Parr's work also seems to echo Sontag (2002), who sees travel as a strategy to accumulate images, and photography as an act of non-intervention. According to Sontag, while the camera offers tourists the opportunity to 'certify experience' (2002, p. 9), it also refuses experience; or at least what would qualify as an 'authentic' experience of the place visited. The only activity Sontag grants to tourists with cameras is a voyeuristic one. As such, tourists only *appear* to participate and they are not actually immersed in the travel experience. Certainly, assuming that actual immersion can only occur in a single, very specific manner and that looking through the camera can never be an actual experience makes

Sontag's claims about the quality and type of tourist experiences at least problematic. Nonetheless, her claims are important for they cast doubt on the camera as a tool of empowerment as much as they challenge one's ways of looking and the reproduction of stereotypical photographic representations.

Taylor (1994) creates a hierarchy in touring by distinguishing between travellers, tourists and trippers. Travellers, with more time and money on their hands, are considered to be more serious and dedicated explorers. Tourists set out to confirm what they already know through promotional material, while trippers are thought of as having the most superficial interest in their destination. Furthermore, Taylor prescribes different ways of seeing for these three groups: 'Travellers practice the gaze, which is contemplative and penetrative; tourists glance, which is accumulative but shallow; and trippers see everything (if they see at all) in disconnected blinks, blurs or "snaps"' (1994, p. 14). The majority, if not all, of Sontag's, Taylor's and Parr's tourists are seen as mindlessly recording what they see without being truly immersed in unfamiliar experiences. The camera functions as a filter, or as a transparent wall, which separates the known from the unknown, offering a sense of control and security. None of Parr's tourists could be considered a 'contemplative traveller'.

Furthermore, in many of Parr's photographs the host and the tourist are often represented as separated by economic power, nationality or social class. In one of his photographs for instance, we see a few Indian women and children in traditional costumes closely surrounding a young white woman in tight pink shorts, seeking her attention in order to sell their colourful belts (Parr, 2007b, p. 39). The young woman looks at the merchandise in contemplation, a few belts hang on her shoulder, yet she remains distant from the local crowd. There is no indication of contact between her and the rest of the women and children, while the background scenery, palm trees and a bright blue sky in the far distance, remind one that she is only there for leisure. This photograph negotiates issues of power and points out the sustained division between developed and underdeveloped countries, even within the framework of globalism. In the photograph, both the belts and the women/children are objectified: they are both part of a marketable image of an authentic, local scene.

The Tourist Stage

Small World offers a view not only of tourist behaviour but also of the tourist stage. In Bethlehem, a seemingly content-looking man in a white keffiyeh on his head and a belt buckle that says 'Jerusalem', sits outside a souvenir shop that sells nativity scenes. On the shop door hang two signs and a crown of thorns (Parr, 2007b, p. 46). The signs read 'CHRISTMAS CARDS' and 'Lufthansa'. One can imagine this scene framed differently. Many tourists or commercial photographers may have taken a close-up of the old man in his white headdress. His ageing dark features, the small moustache and his clothing would have made a picturesque snapshot of the eastern native host, a pleasant souvenir of a truly authentic experience. The wider-angle scene, which includes the signs on the door and the visual proofs that this man is sitting outside a souvenir shop, would have detracted from the 'authenticity' of the photograph. By ignoring the mundane or unattractive, both promotional imagery and vernacular tourist photography seem to reinforce an idealised version of tourism and the myth of a perfect vacation (Markwell, 1997). The work of Martin Parr undermines and plays upon these dominant visual representations. In effect, through his photographic choices, Parr presents this scene as an example of the artificial tourism stage and his work seems to maintain Boorstin's claim that:

> the tourist sees less of the country than of its tourist attractions. Today what he sees is seldom the living culture, but usually specimens collected and embalmed especially for him, or attractions specially staged for him: proved specimens of the artificial. (1987 [1961], p. 102)

Tourists are often presented with a 'staged authenticity' by the host: an environment that appears to be authentic, but which is in reality well-orchestrated (MacCannell, 1999 [1973]) to lure the tourist by satisfying those exact visual stereotypes sustained by both the media and tourist photography. The man in the photograph described above can easily be considered a tourist attraction, a staged specimen of the artificial. Interestingly, though, the objects surrounding him clash with this well-orchestrated setting that aims to create the illusion of authenticity. They are objects that are well recognisable by the global tourist. Both Christmas and Lufthansa are signs that anyone can relate to, and in effect, they seem to become a link between the known and the unknown and a means of access to the culture visited.

Parr also notices tourism conventions created by the host to facilitate tourism. In one photograph, Parr presents a restaurant sign from Playa de Las Americas in Tenerife, a popular holiday destination in Europe (2007b, p. 90). The photograph is of a set of dishes offered in the restaurant's menu; most of them a combination of eggs, beans and French fries, lying on a green and white chequered tablecloth. Next to each plate there is a clear indication of the price. This arrangement might look unappetising but is immediately understood by most tourists regardless of nationality and it reminds us of many similar labels found in touristic destinations around the world. All tourists need to do in these cases is to point to the picture of the food they want without uttering a word in a foreign language and Parr seems to be acutely aware of such tourist conventions.

The Artist as a Traveller–Critic

As shown in the previous sections, Parr's photography problematises the tourist gaze, the artificial tourist stage and those conventions that aim to facilitate tourism. It becomes a critique of the global, uninterested tourist, and in so doing highlights modernist views of tourism that began with Boorstin. According to Dyer:

> There is no way around it: to travel, either as backpacker or package tourist, is to be forced into being an incessant consumer. Factor in queues, hassle, jetlag and tummy upsets and it's a wonder, even now, when travel has become so *easy*, that

> people still want to do it ... That's what the pictures in *Small World* depict: the form and state of modern, faithless pilgrimage. (2007, p. 7, emphasis in original)

Perhaps because Martin Parr is documenting the tourist as an 'incessant consumer' he is considered among those 'oppositional photographers' who engage in a wider debate about the tourism industry. Taylor explains:

> Oppositional photographers rework the language of tourism, and play with its appearance and concerns. They take part in a constant dialogue with the everyday practices of tourism and with its standard imagery. Their success depends upon their ability to embrace the cliché and alter it, making it appear new in ways which are surprising and not necessarily subtle. (1994, p. 241)

For Taylor, Parr 'remains outside the group [tourists] in the privileged position of the observer. This position establishes him as a traveller, looking with greater authority than trippers can' (1994, p. 243). Keeping in mind Taylor's hierarchy of travellers, tourists and trippers, Parr seems to adopt the most advantageous position of the traveller who gazes upon tourists and trippers. He acts as a *traveller–critic* who adopts the rhetoric of globalisation in order to critique contemporary consumption and reveal the discrepancies between promotional material and reality.

Furthermore, according to Martin Parr's website, Parr highlights the bizarre yet widely accepted tourist behaviour in order to create 'symbols that will help future generations to understand our cultural peculiarities' (Parr, 2013). This comment implies that people are currently too immersed in these cultural peculiarities to acknowledge them or question them while Parr plays the role of a social critic who makes them visible. However, as revealed later in this chapter, this is not necessarily the case since tourists also notice and even photograph these cultural peculiarities.

Postmodernist View of Parr's Tourists

The new tourist conceptualisation focuses on the individual tourist and recognises that his/her behaviour, or better 'choreographies', is influenced by complex social and cultural relationships. Tourists are not just extras in a tourist stage but they unavoidably shape it since they bring with them 'their gendered, aged, sexed and racialized bodies' (Haldrup and Larsen, 2006, p. 279), as well as their complex personal and group relationships that unavoidably creep into their photographs. As van de Ven describes her experience as a tourist in Kolkata, 'the camera does not gaze; instead, it creates a stage around itself: a stage that includes the person holding the camera' (2011, p. 143). In effect, what was initially read as a criticism of inauthentic travelling in Parr's photographs of the tourism stage, can now be read as a document of choreographies by tourists who themselves (re)create the same stage.

In addition, one should take into consideration that the memories triggered by travelling and the uses of the photographs one takes during a journey are not specific to one particular time. Instead, both memories and uses extend in time. For example, tourists might partially experience a destination through camera lenses, but they feel compensated since they use photography after their trips in order to share experiences with loved ones, create narratives, trigger memories, and further social engagement. By extending beyond the boundaries of the before, during, and after of the travel experience (Scarles, 2009), tourists' photographs inform identities by 'potentially collecting illustrations and titles for a self-narrative' (Belk and Yeh, 2011, p. 349). Parr's tourists, who spend their time behind their camera lenses, in reality are not just reproducing specific scenes they have seen in promotional material. Instead, they are constructing images of self, demonstrating their dreams and achievements, and preserving relationships and experiences of which they are proud of for future contemplation and sharing.

As van de Ven (2011) argues, the tourist might indeed initially visit a place attempting to confirm those stereotypical images already prescribed in his/her mind, by means of photography. But soon it is possible to see a place and its people as part of a much more complex reality. He asserts, 'Without a camera, we can look away and walk away from any challenges to our preconceptions, but the camera stops us in our tracks' (van de Ven, 2011, p. 149). This is an idea that contrasts completely with Sontag's previous notions of the camera as an obstacle to looking and experiencing a location. Instead, it acknowledges the complexities of tourist practices and the failure of the modernist notion of the passive gaze to deeply analyse tourist photography.

Similarly, Parr's modernist description as the visionary traveller–critic who is set apart from common tourists and trippers, unveiling through his work what is not obvious to others, cannot but be further questioned. This point of view might initially support Boorstin's and Taylor's contrived (and largely outdated) distinction between travellers and tourists, but upon closer inspection it simply reinforces one's suspicions that such distinctions are indeed problematic. For if one accepts such distinctions, then Parr's seeming distance from the ordinary tourist does not necessarily mean that he is a traveller who engages with what he sees in an objective and meaningful manner. Since Parr simply records behind his camera lens, guided by his own assumptions about tourism, he can be equally considered as alienated from the people and locations he photographs as any other *tourist*.

Indeed, just like tourist photography, Parr's photography seems to document with incredible acuteness the visible (stage, behaviour) at the expense of the implied (meaning, relationships). Perhaps for this reason his work has been criticised for using his subjects as stooges for his advertising-like, ironic images that erase any sense of reflective subjectivity or an active conception of audience (Dawber, 2004). According to Dawber:

> Parr's failure is not his relentlessly upbraiding demeanour but his refusal of the (well-worn) impasse of naturalized documentary: its voyeuristic rendition of surface, rather than substance, of class relations; its obliteration of the photographer's own socially grounded subjectivity; its positivistic bias towards the empirical world; its denial of praxis. (2004, p. 258)

This description of voyeuristic and surface-oriented photography could easily be applied to tourists' own photographs if one follows the modernist analysis.

Furthermore, Forbes mentions that Parr's work might have generated critical insight into today's tourism but 'on the whole it does little to mobilise itself against the culture of the spectacle' (2001, p. 233). In effect, Forbes questions whether or not Parr's comic but harsh revelations actually play a role in changing what he is criticising. However, this criticism should not compel us to dismiss Parr's work as an 'aimless, contemporary comedy of manners' (Dawber, 2004, p. 258). Parr's work remains fresh and relevant because through his selective, mirror-like reflections of the tourism world we can recognise ourselves as tourists. Parr offers a self-reflective, ironic stance that might be adopted by more tourists than one might think.

The Artist as a Professional Post-tourist

Most tourists are aware of the tourism stage, the inauthenticity of the tourist experience and that they are willing players in a series of games (Feifer, 1985). Fiefer (1985) introduced in 1985 the concept of the *post-tourist.* Post-tourists are not only aware of the construction of the stage, but are also aware of the commoditisation of place and self in a playful and sometimes ironic manner (Feifer, 1985). One of the characteristics of post-tourists is being reflective on their own practices and experiences. According to Edensor:

> … another form of partially improvised performance is that of the 'post-tourist' (Feifer, 1985; Urry, 1990). Acutely self-reflective, and predisposed to distantiation from the codes of performance, such actors play with normative scripts and roles, mocking the constructed and regulated nature of such performative codes while ironically intimating their awareness of the restrictions of the role. (2000, p. 56)

Playing with normative scripts, mocking the constructed codes of tourism and being ironic is what Martin Parr does best. In fact, this is what renders him the ultimate post-tourist.

However, Parr is not the only tourist who photographically reflects on tourism practices. Another example of tourists' practices of self-reflection is the online photography group 'Pisa Pushers' on the popular photo-sharing application Flickr (Flickr, 2013). This Flickr group features photographs taken by tourists at the Leaning Tower of Pisa showing *other* tourists 'pushing' it. The group was initiated in 1995 by Magnus Wrenninge, a visual-effects artist, the same year that Martin Parr's first edition of *Small World* was published. Even though Wrenninge, an amateur photographer at the time, was not aware of Parr's work, he seems to have had similar reflections as Parr. In an e-mail interview Wrenninge mentions:

> I remember finding it absurd how many people wanted to have their picture taken pretending to support the tower. It just seemed like a natural thing to take

> pictures of them, expressing the absurdity by changing the point of view and the context. (e-mail correspondence, 3 Oct. 2012)

Furthermore, Wrenninge explains that he noticed other tourists taking similar pictures of other tourists and thus he does not consider his photographs as unique. Although the contributors to the 'Pisa Pushers' are mostly people who have been at the location as tourists and not as artists engaged in a photographic project, their photographs are not that dramatically different in subject matter and style from Parr's photograph of the same location (see for example Figure 11.1). Both Parr and the contributors to this Flickr group capture the behaviour of other tourists and the ways in which they engage with both the location and other visitors in a playful dialogue, two characteristic elements of the 'post-tourist'.

Indeed, another way to envision the role of Martin Parr is to see him as a *professional post-tourist* who is not simply recording in order to offer a critique but who also reflects on his own tourism practices along with the practices of his fellow tourists in a systematic and persistent way. As Nicholson says of the people featured in Parr's photographs:

Figure 11.1 Fotobananas

Source: Flickr group 'Pisa Pushers', used with permission.

> We're certainly laughing at the people rather than with them, and yet there isn't anything especially cruel about the photographs. And, for some of us at least, there's the laughter of uncomfortable recognition. (2002, p. 87)

One could claim that Parr does not necessarily adopt a superior position in relation to his subjects and that the 'laughter of uncomfortable recognition' was also experienced by the artist himself. After all, from the British middle classes to self-portraits and tourism in general, Parr's themes are usually associated with everyday life.

Conclusions

One cannot deny that Parr's reflections on tourism are highly specific. The consistent adoption of the same formal elements (wide angled scenes, ring flash) and the repetition of the themes and issues he wishes to reinforce with his work are indicative of his personal convictions as well as his experiences of tourism. By placing the emphasis on combating promotional myths and pointing out the bizarre, Martin Parr constructs his own 'propaganda', rendering the tourism world a *small repetitive one* where all tourists behave in the same, often comical, manner, and in which the landscape is adapted to serve their needs. From the point of view of modernism, Parr becomes a traveller–critic and his work exposes the tourism stage and the peculiar behaviour of tourists rather than engaging with the individual tourist and his/her meaning making processes.

But this is a very limited way of reading Parr's work. Parr is part of this discourse as much as the viewers and the tourists he photographs. He also travels in order to take his photographs, has become part of the crowds that he follows, and without a doubt has missed a flight, waited in airports, and had lunch in a restaurant and been forced to order a meal off a picture menu because he did not speak the language. Postmodern theories and the notion of the performative turn in tourism studies can shed new light on Parr's photographic practices. A new light that highlights artistic practices which are less hierarchical and more attuned to the artist being part of the discourse with which he engages.

Thus, what is suggested here is a new understanding of Martin Parr's work as a professional post-tourist *in addition* to his role as a traveller–critic. Parr's photographs indeed counter the propaganda power of promotional material and often stand as a critique against mass globalised tourism. At the same time, however, they are an honest, systematic visual narrative showing the minute details of the tourism experience the artist shares with millions of other tourists. In this way, his work moves beyond the superficial modernist readings of the tourist as a passive consumer of inauthentic tourism stages and enters a more postmodern reading in which the self-aware tourist also becomes a producer of a polysemic tourism stage.

References

Bærenholdt, O.J., Haldrup, M., Larsen, J., and Urry, J. (2007). *Performing Tourist Places*. Hants, Burlington: Ashgate.

Belk, R. and Yeh, H. (2011). Tourist Photographs: Signs of Self. *International Journal of Culture, Tourism and Hospitality Research*, 5(4), 345–353.

Boorstin, D. (1987 [1961]). *The Image: A Guide to Pseudo-Events in America*. New York: Atheneum.

Crouch, D. and Lübbren, N. (2003). Introduction, in D. Crouch and N. Lübbren (eds) *Visual Culture and Tourism* (pp. 1–20). Oxford, New York: Berg.

Dawber, S. (2004). Martin Parr's Suburban Vision. *Third Text*, 18(3), 251–262.

Dyer, G. (2007). Introduction, in M. Parr, *Small World* (pp. 4–7). Stockport: Dewi Lewis Publishing.

Edensor, T. (2000). Staging Tourism: Tourists as Performers. *Annals of Tourism Research*, 27(2), 322–344.

Fiefer, M. (1985). *Going Places*. London: Macmillan.

Flickr (2013). Pisa Pushers. Retrieved 12 February 2013, from <http://www.flickr.com/search/?q=pisa+pushers&f=hp>.

Forbes, D. (2001). Magnum's Frame. *History Workshop Journal*, 52(Autumn), 229–235.

Haldrup, M. and Larsen, J. (2006). Material Cultures of Tourism. *Leisure Studies*, 25(3), 275–289.

Haldrup, M. and Larsen, J. (2010). Performing Tourist, Performing the Orient, in M. Haldrup and J. Larsen (eds), *Tourism. Performance and the Everyday: Consuming the Orient* (pp. 1–19). London, New York: Routledge.

Kaplan, C. (2005). *Questions of Travel: Postmodern Discourses of Displacement* (2nd edition). Durham and London: Duke University Press.

Lane, G. (2006). Photography from the Photographer's Viewpoint. *The Art Book*, 13(4), 15–16.

Larsen, J. (2005). Families Seen Sightseeing: Performativity of Tourist Photography. *Space and Culture*, 8(4), 416–434.

MacCannell, D. (1999 [1973]). *The Tourist: A New Theory of the Leisure Class*. Berkley, Los Angeles, London: University of California Press.

Markwell, K. W. (1997). Dimensions of Photography in a Nature-based Tour. *Annals of Tourism Research*, 24(1), 131–155.

Nicholson, G. (2002). Objects of Derision. *Modern Painters*, 15(2), 86–91.

Osborne, P. (2000). *Travelling Light: Photography, Travel and Visual Culture*. Manchester: Manchester University Press.

Parr, M. (2006). *Tutta Roma*. London: ROCKET.

Parr, M. (2007a). *Mexico*. Blue Jay, CA: Hawking Books.

Parr, M. (2007b). *Small World*. Stockport: Dewi Lewis Publishing.

Parr, M. (2010). *Machu Picchu*. Portland, OR: Nazraeli Press.

Parr, M. (2011). *Japan*. Tokyo: Super Labo.

Parr, M. (2013). Martin Parr. Retrieved 12 February, from <www.martinparr.com>.

Robinson, M. and Picard, D. (2009). Moments, Magic and Memories: Photographing Tourists, Tourist Photographs and Making Worlds, in M. Robinson and D. Picard (eds), *The Framed World: Tourism, Tourists and Photography* (pp.1–37). Surrey, Burlington: Ashgate.

Romig, R. (2010, April 8). Off the Shelf: Martin Parr's 'Small World' [The New Yorker Online Blog]. Retrieved 12 February 2013, from <http://www.newyorker.com/online/blogs/photobooth/2010/04/martin-parr-small-world.html#slide_ss_0=1>.

Scarles, C. (2009). Becoming Tourist: Renegotiating the Visual in the Tourist Experience. *Environment and Planning D: Society and Space*, 27(3), 465–488.

Sontag, S. (2002). *On Photography*. London: Penguin Books.

Stylianou-Lambert, T. (2012). Tourists with Cameras: Reproducing or Producing? *Annals of Tourism Research*, 39(4), 1817–1838.

Taylor, J. (1994). *A Dream of England: Landscape, Photography and the Tourist's Imagination*. Manchester, New York: Manchester University Press.

Uriely, N. (2005). Theories of Modern and Postmodern Tourism. *Annals of Tourism Research*, 24(4), 982–984.

Urry, J. (1990). *The Tourist Gaze*. Los Angeles, London, New Delhi, Singapore, Washington DC: SAGE.

Urry, J. and Larsen, J. (2011). *The Tourist Gaze 3.0*. Los Angeles, London, New Delhi, Singapore, Washington DC: SAGE.

van de Ven, A. (2011). The Eyes of the Street Look Back: In Kolkata with a Camera Around my Neck. *Photographies*, 4(2), 139–155.

Weski, T. (2013). Martin Parr: Introduction. Retrieved 12 February 2013, from <www.martinparr.com>.

Chapter 12
Hunting the Touristic Animal through Ceramic Art in Alberta, Canada

Fiona Thompson and Kevin Hannam

Introduction

This chapter is based upon collaboration between a ceramic artist and a tourism academic and builds upon previous work that has examined the relations between tourism and ceramic art (Thompson, Hannam, and Petrie, 2012). With its central focus on artistic practice, this chapter offers an original and creative interpretation of taxidermic collections in Alberta, Canada through ceramic artworks. A dialogue between art and travel experiences of the artist, Fiona Thompson, in the province is mapped. As a collaborative project, the chapter discusses Thompson's responses to source material gathered from Alberta, the exploration of particular themes and the subsequent development of artwork. It also explores the engagement of the work with broader theories, in particular relating to hunting as a touristic activity in the region. To begin with we contextualise this artistic project with a review of existing academic literature on animals and tourism and hunting as a leisure pursuit in particular. Subsequently we go on to discuss taxidermy and the culture of collecting from a historical and scientific perspective. Finally, we also examine the touristic context of Alberta, Canada and the ambivalence of art and souvenirs. In terms of methodology the chapter outlines the source material that was collected and the overall 'journey of work' undertaken to create specific artworks that are then evaluated. This evaluation focuses upon a collection of ceramic plates that were created by the lead author of this chapter.

Animals, Tourism and Hunting

The relations between animals, tourism and hunting are multi-faceted. In the media humans are frequently portrayed as environmentally destructive with wild animals under threat. Moreover, postmodern human relations with animals are arguably characterised by a stronger emotional and moral content, a greater zoological range of involvement but nevertheless requiring demands for more regulation and order (Franklin, 1999). For example, the performance of a species in the wild for tourists is not reliable, nature does not perform at the right time, but nevertheless this expectation has to be 'managed'. However, the notion of the timing of 'wild' events is significant because if nature is manipulated spatially or temporally then the sight loses its authenticity and value for the eco-tourist (for he or she might as well go to the zoo). Thus the tourist is left at the mercy of nature for a performance, unless the tourist has sufficient time and money to spend a lengthy period looking for certain species. Thus, surveillance of species as a tourist experience becomes important in order to verify the knowledge that they are available for touristic consumption: that there is a chance of an encounter. If the chance falls too low tourists may go elsewhere (Hannam, 2004; Hannam and Knox, 2010).

Moreover, in one study it was found that postmodern western ecotourists tend to anthropomorphise animals (Russell, 1995). Similarly, cute and cuddly animals are used for tourism marketing purposes. It has also been argued that these species will generate better public response and in the process of protecting one such species, such as pandas, countless others will also benefit. This trickledown effect, however, has been criticised by others who argue that it marginalises some animals in favour of others, and fails to persuade the potential tourist to view nature as a total ecosystem (Russell, 1995). Furthermore, Beardsworth and Bryman (2001) have argued that there are four 'modes of engagement' through which wild animals are experienced by tourists. Firstly, they note the physical 'encounter' with an animal in its own environment, a relatively unusual experience confined to those who actively seek such encounters through tourism. This may be consumptive in terms of hunting, fishing or shooting or non-consumptive in terms of photography (Lovelock, 2008). Secondly, we have the 'representation' of wild animals which, of course, has a long history in books, on television and in films and cartoons. Beardsworth and Bryman note in this context that:

> … where cartoon representations of wild animals are involved, any links with the appearance and behaviour of actual species may become extremely attenuated. However, even though cartoon animals are typically given human attributes (e.g. bipedal walking, speech, a sense of humour), this anthropomorphism does not necessarily prevent their being used by audiences as sources of 'knowledge' about the real thing. … Within the Disney idiom, there are powerful currents of anthropomorphism and sentimentality in the depiction of animals (both in live action and cartoon contexts), and these currents themselves become incorporated into the virtual capital of consumers of Disney cultural and commercial products. (2001, p. 86 and p. 99)

Thirdly, we have the direct 'presentation' of wild animals in captive settings where they are subject to scrutiny. This also clearly ties in with Norton's (1996) research on safari tourism in Africa that found that some tourists felt that their experience had become 'zoo-like' due to the congestion of too many jeeps in the park. Fourthly, Beardsworth and Bryman (2001) argue that we also have the 'quasification' of wild animals through the creation of impressive fake artefacts such as animatronic animals in museums or even the cuddly soft toy animals that we see in the shops. To this, we might add a fifth mode of engagement, that of the touristic animal in artistic practice which draws upon and is inspired by the previous modes of engagement. In particular we discuss how artwork draws upon methods of taxidermy and collecting below. We then discuss the empirical research context of Alberta, Canada.

Taxidermy and Collecting

Taxidermy derives from the Greek taxis, meaning arrangement, and derma, meaning skin, and specifically refers to the practice of re-creating an animal, or to take it literally, rearranging skin. It is an act of preservation. As summed up by Madden: 'In taxidermy, the empty becomes whole and what was dead becomes lifelike …' (2011, p. 2). As a practice it has been around at least since the 1600s, possibly earlier. However most specimens pre-1740 have decayed and disintegrated and there is apparently little evidence of early specimens (Madden, 2011). Now an established industry, particularly in North America, it is closely linked with the hunting industry, with the act of collecting and public or personal display of such collections. Natural history museums, the home of much taxidermy, and the study of natural history, became popular by the late 1800s. Barber (1980) investigated the rise of this popularity within the context of social structures and scientific developments such as Linnaeus's system of classification. Collections were developed by individuals with an enthusiasm for the natural world and scientific understanding. Charles Waterton, a nineteenth-century explorer and naturalist, for example was travelling, hunting, killing, preserving and creating displays of exotic animals at Walton Hall (Caroll, 2004). Walter Rothschild was another well-known British example of another nineteenth-century zoological collector who opened a private museum (Rothschild, 2008). The act of collecting can also be associated with obsession; the urge to continually add to a collection. When there is no maximum number of objects/ specimens available to collect, when is a collection complete? For further details on collecting in general, Pearce offers much insight and analysis into the field, how collections may be interpreted and the motivations to collect (Pearce, 1994, 1998). For Baudrillard (1994), collecting is about passion, possession and fanaticism. We now examine this in the empirical context of Alberta, Canada.

Alberta, Canada

Within the province of Alberta, the diverse landscape, its native wildlife and outdoor pursuits such as hunting and hiking make up a substantial part of the touristic 'package'. Animals such as the bison or buffalo, deer and mountain goat act as a metonym for Alberta. They are often strongly associated with the Native peoples and the history of the region (see Figure 12.1). Their images feature heavily on souvenirs, are romanticised through calendars, postcards, toys and other objects, and may be used as part of the touristic promotion of the area. The website www.travelalberta.com for example features wildlife viewing, hunting and fishing as tourist activities.

Figure 12.1 Display at the 'Indian village' attraction, Calgary stampede 2008
Source: Artwork and photo by Fiona Thompson.

Moreover, the stuffed and mounted animal is found widely throughout the region. Displays of animals can be viewed in a variety of contexts: hotels, interpretive centres, independent stores, bars and shopping malls. Animals here are subjected to the tourists' gaze with varying degrees of prominence and experiencing these taxidermied objects is predominantly visual in nature. They invite visual consumption, and in certain cases physical interaction through touch. For those behind glass, there is even more emphasis on observation through a pictorial 'frame'. Simultaneously the glass invites the viewer in whilst also creating a physical barrier and therefore creating further separation. These animal collections, embedded within the touristic language of this region, confirm for the visitor to Alberta that they are 'here'.

As a visitor to the area, and importantly as an artist, everything has potential for visual consumption (Thompson, 2010). Theoretically, the artwork discussed in this chapter engages with the recent tourism research on hunting and tourism, especially Lovelock (2008), Hannam (2008), Foote and Wenzel (2008), and Campbell (2008). The dioramas and displays may narrate previous hunting successes, and in other cases are used for commercial purposes. These displays have potential for multiple interpretations however, and like many other touristic sights, their presence is complex, concealing many issues. Rooted in a previous artistic research by Fiona

Thompson, this project draws upon earlier artworks informed by a previous field trip to Canada. Many locations and particular sights were revisited, in addition to new ones during the most recent trip in 2011. This revisiting offered an opportunity to reflect on previous interpretations, whilst approaching the material from a new viewpoint. Paradoxically by becoming more familiar with the region, the differences between this region and 'home' were further highlighted (particularly as the focus was on a potentially provocative subject matter). This sense of ambivalence is present in much of Thompson's artworks.

Methodology: Source Material

Much source material for the artwork discussed in this chapter was acquired in a large outdoor accessories store; part of the North American chain 'Bass Pro Shops'. Detailed large scale dioramas of native wildlife provided an environment in which to market products, many of which related to hunting and fishing (see Figure 12.2 below). Evidence, through informal discussion with local residents of Calgary, indicated that many of these exhibits were constructed from several animals to create a whole. For the casual viewer however this is not clear; instead there is an assumption of each being a complete animal. The complexity and visual qualities of the Bass Pro Shop displays could be compared to that of a

Figure 12.2 Display at Bass Pro shop, Calgary 2011

Source: Artwork and photo by Fiona Thompson.

museum natural history collection. Yet the motivation is somewhat different. There is a complexity of meaning embedded within this kind of sight; unlike the museum collection these 'specimens' are constructed purely for commercial gain. They are open to interpretation and extremes of views; they could be seen as humorous, as kitsch, as unethical, or as a goal for the hunter. Yet, they are metonyms of Alberta, and of a popular pastime. Through the veneer of the attractive display the realities of hunting are perhaps given a positive spin. A dilemma is presented; should the viewer disapprove, or accept that hunting and the role of the animal as display is integral to the region? In addition it raises the question of whether it is appropriate for the visitor to criticise the activities of other places. It is precisely this dissonance or ambivalence that is of interest and that gives the collections as a subject matter much potential for analysis and creative interpretation.

In the animal displays the focus is on what is implied; giving the impression of reality is central to the quality of an exhibit. The audience is not led to believe that the dioramas are real; it is the representation of reality that is key. The carefully constructed settings are in reality only a veneer, but this is all that is required. Staging of the animals in their natural habitat creates a vista of the landscape; a peculiar, scaled down self-contained stage set of the view that does not move or change. The 'great outdoors' is brought indoors in a safe and controlled setting. Both the animals and their three dimensional backdrop are fixed permanently in place, rendered into naturalistic poses. The notion of 'photographic seeing' (Sontag, 1977) becomes significant. These displays are arguably composed to reflect the view through the camera lens (and perhaps are based on photographic compositions), and thus to create a visually pleasing vista. It reinforces Sontag's argument that we are conditioned to see from a particular perspective.

Other locations of significance included a museum of native wildlife (The Den) located in the basement of a hotel in Jasper National Park. This collection was first amassed in the 1960s by a hunter, and sold to its current owner who set up the museum. The animals were divided into groups and 'families', placed amongst fake rocks, trees and other naturalistic elements. Where only the heads of particular animals had been preserved, they were placed to give the impression of emerging from undergrowth, trees or caves. Like many other collections, this had the potential to trigger a sense of nostalgia and sentiment, although also perhaps a sense of unease. Unlike the Bass Pro Shops displays however, there is arguably less ethical conflict. Although theoretically the same; the distance of time, context and intent renders these animals more 'acceptable'. It is a small, privately owned collection, and in fact could be considered an authentic attraction in the sense that it is unique and specifically related to the region.

The Native peoples' Interpretive Centre 'Head-Smashed-In-Buffalo Jump', meanwhile, used native wildlife in a story telling context. The physicality of a genuine buffalo for example, gave this story depth and a sense of conviction. Again we return to the notion of the authentic; for the tourist, a sense of being immersed into past cultures was a key characteristic of this Centre. The opportunity to touch and see the animals brought the experience closer to a 'genuine' experience.

Native peoples working at the Centre used particular examples of stuffed wildlife to illustrate the story telling element of an interactive display. The placing of the animals could be deemed more appropriate and more ethical in this context than the Bass Pro shop for example.

The Plate: Language and Context

To create a context for the artwork itself, we will first offer a brief background to the commemorative vessel. There is a rich history of the ceramic vessel as a vehicle for commemoration and communication which certainly impacts, directly or indirectly, on contemporary ceramic practice. Many of the vessel forms still explored today have their roots in historic objects (see for example Godden (1985, 2004), Drakard and Holdway (2002) and Baker (1984) for further discussion on the history of printed ceramic vessels). The ubiquitous plate is not a new object; rather it has been used for centuries for utility, ornament or to commemorate events. A detailed study of ceramic history is out of the scope of this chapter, we will therefore instead briefly focus on a small sample of Canadian-related plates to create a context. Firstly a mass produced plate (made in Japan, date unknown) commemorating the town of Edmonton, Alberta, on which is illustrated a collage of popular landmarks (see Figure 12.3 below).

Figure 12.3 Commemorative plate

Source: Photo by Fiona Thompson.

The images are originally drawn but reproduced using typical ceramic factory production methods. Its flat simple format makes it an idea canvas. The plate as a form is unspecific; it does not have an affinity with particular locations, cultures or period in history. It is the text and image that locates this plate to the town (and paradoxically to its country of manufacture). This characteristic makes it ideal for its purpose.

Secondly, a plate, also domestic in scale, features a single photographic image of Lake Louise, Alberta. Manufactured in the UK by Royal Doulton, it dates from the 1960s from the 'Series Ware' edition of commemorative pieces. It is simpler and perhaps more tasteful than the Edmonton plate, but both use the plate as a canvas for commemorating an enjoyment and fascination with 'other' places. Medalta Pottery, Medicine Hat, Alberta (now a museum and artist-in-residence centre), contained many examples of commemorative ware made during the late twentieth century. Particularly of interest to this project were plates commemorating local racehorse winners, and scenes of the local landscape and wildlife. This specification of place, time and event gives a presence to the objects and with progression of time, sentiment and nostalgia increases.

It may seem that these mass produced examples are far removed from contemporary artistic interpretations of the plate, and yet they have many similar themes and purposes. The contemporary ceramic artist can use it as a vehicle for exploring ideas and creating imaginative interpretations of the plate (see Scott, 2002, 2007 and Hare Duke and Theophilus, 2011). Scott and Bennett (1996) offer further examples of contemporary and innovative approaches. It is precisely this familiarity and its simplicity that made it appropriate for this body of artwork. Its form offers a perfect canvas for imagery and concept, whilst retaining a link with ceramic traditions and language. However, there can also be a difficulty in approaching a form that has been made and remade countless times. For the artist, the question often arises of how to create something new. Individuality and innovation can be achieved however through a personal approach to interpretation and communication of subject matter.

The Artistic Journey

The process of observing, selecting, acquiring and interpreting source material[1] and the translation into concrete ideas provided a basis for making the artwork itself. This transition or journey from experience/observation to resolved artworks is as much part of the work as the physical making processes. Relevant literature had varying degrees of influence, both prior to and following the field trip; for example underpinning the interpretation and use of the photographic image (Sontag, 1977). Once the source material had been initially viewed and edited, and idea development was underway, contextual literature was sourced that might support and inform this process.

1 Artists' source material may vary widely; in this case it consists of written notes, sketches, photographs and gathered material such as leaflets.

So the artwork begins with touristic experience; embedded in this is the notion of 'photographic seeing' as discussed earlier (Sontag, 1977). As we will see later in the chapter, the photographic image is also a significant part of the artwork itself, both physically and conceptually. For the artist, 'data' gathered is intentionally researcher-biased; for the purposes of this project a variety of animal collections were sought out and documented. The act of travelling through the landscape also provided much material, with photographs enabling vistas to be viewed for longer than in reality.

The process of selection, editing and choosing images for further development occurred over a period of several months following the field trip. It was often revisited several times once the ceramic pieces were under construction; the images become familiar and can offer new insights. Particular images tended to dominate; those with a stronger visual quality for example. Overriding themes of ambivalence, authenticity, representation and commemoration emerged through the analysis of material. Digital manipulation of the selected images took place separately from the construction of three-dimensional form. The relationship between the two, and potential compositions were considered by the artist. The final compositions however are not rigidly predesigned; they developed in various stages with layers of image, colour and text 'collaged' over each other. The underlying layers dictate the placing of appropriate images. This process enables individual approaches to, and juxtaposition of, similar forms and repeated imagery. Repetition is significant to the artwork, reflecting the characteristics of photography and of the taxidermic animal itself. The original photographs are used to create a stencil for the cut out animal silhouettes, and for the stencilled or monoprinted animals, thus referring back to the idea of photographic seeing. The photograph is therefore the basis for the compositions; from that stems various ways of representation. There is also a parallel with the collections of study animals, where numerous specimens of the same animal will be collated to compare the variations within a species.

The Artwork: 'The Albertan Animal'

A collection of plates was created for this project; intended to be viewed as individual objects within an overall single piece. Variations encompass round, oval or rectangular forms with double walls and printed and painted imagery (see Figure 12.4). Images, colour and text were layered up on the surface, playing with scale and context.[2] The overall quality of the surface is smooth, with layers of image and colour flattened and blended. Paradoxically by limiting the form within the series, it allowed for further exploration within it. Like animal species, there are

---

2 The processes of making: sheets of clay are cut out and the blank forms are assembled by hand. Layers of colour and some text are applied, with stencils/masking used for silhouetted shapes. Glazes are applied in another layer, and finally the main images and text. The pieces are fired three times between particular stages. Many test pieces are made before the final finished objects, resolving composition, form and finer details.

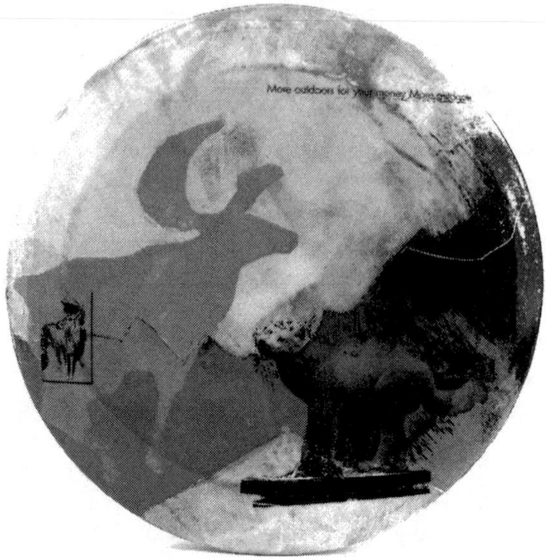

Figure 12.4 Plate from 'Albertan Animal series' 1

Source: Artwork by Fiona Thompson, photo by John McKenzie, used with permission.

multiple variations within one type, allowing for a deeper theoretical exploration of text and context. The plate itself is of course a symbol of domesticity; whilst often being used as a souvenir. As previously mentioned, it is universal, and found in multiple forms, sizes and contexts. It was chosen for this reason as an object for commemoration, and as it offers a broad flat surface for imagery that is 'more than representational' (see Anderson and Harrison, 2010).

The animal as subject is firmly established within art generally, and in a ceramic context. Artistic interpretations vary; illustrative or abstracted in two dimensions, photographic and physical interaction through the use of actual taxidermy. Poliquin (2010) offers some interesting perspectives on the world of taxidermic art, particularly considering the implications of context on how it may be read. She states: 'Taxidermy has become a potent medium to discuss a variety of pressing issues: the contours of the line between humans and other animals, questions about conservation and species loss, and more basically to provoke deeply enigmatic encounters with the natural world. The thing is recognised as an animal, but the nature of the encounter is less clear' (Poliquin, 2010, p. 1).

This artwork however focused specifically on a touristic context and representation of taxidermy. It looks at the preserved and collected (and in some cases commercialised) animal as a symbol of the region; the representation rather than the real thing. The work is concerned with playing with sentiment and assumptions, and in part with traditional presentations of animal imagery in ceramic. As in hybrid geographies the complexity of the animal as subject matter is hidden in these plates (Whatmore, 2002). Instead they are presented as elevated commemorated creatures, with subtle references to their darker context. They are romanticised; fixed permanently in place on ceramic surfaces. The cut outs, like with earlier work, presents the space of an animal that is no longer there (see Figure 12.5). They exist as images, objectified and pinned down in a two dimensional photograph.

The landscapes of the region are diverse and extreme; with large distances between towns. Prairie, mountains and forests are very much embedded into the Canadian 'tourist-scape' and national identity (Osborne, 2001). For the artwork, these were an appropriate choice for backdrop to the animals' imagery, both visually and metaphorically. The landscapes have been interpreted as slightly abstracted blocks of colour, with incised lines denoting the edge between sky and mountain range for example. Layers of colour applied by brush loosely reflect colours of the landscapes. Photographs and drawings were used as a reference point, again to retain that link with the photographic composition and to refer to specific views seen. The scenery was interpreted as visual backdrop, and in many ways was experienced as such during the field trip (as opposed to physical engagement through hiking for example). The framing of the plates contains, miniaturises and in a sense tames the scenic vista, rendering it portable on a domestic scale. The landscape could be seen in this context as a product, as a decorative backdrop. In a sense it is packaged as a touristic product, particularly within the boundaries of the national parks. Creating

Figure 12.5 Plate from 'Albertan animal series' 2

Source: Artwork by Fiona Thompson, photo by John McKenzie, used with permission.

a form (built by hand) that is hollow allowed for several pieces to reveal the layer behind; a silhouette of animals cut through the top layer. It makes reference to the fact that the 'real' animals are hollow; only skins or surface remains to give the illusion of reality. Displacing geography, the once animate is now inanimate. The negative space has connotations of the empty, or missing. A variation of this is the pieces that are completely enclosed, with only the surface available to view. So like the animal displays, there is nothing to see behind the veneer. Three dimensions are reduced to a flat plane, with the animals represented in different ways. Stencilled silhouetted blocks of colour could be read as either negative or positive space; mono-printed outlines or altered photographic images represent the animals in another variation. Many of the images (the photographic transfers) are cropped, altered and the original backgrounds partly removed using photo imaging software. Adding another layer of representation, some of the images are photographs of illustrations viewed in various locations. In reference to the Bass Pro animals, some of the images are de-constructed into separate parts. They are partly translucent, revealing the layers of colour and text underneath. Text on the pieces is taken from various sources. It may make reference to touristic language (Dann, 1996) such as the phrase 'Welcome to Jasper'. Other snippets of text are taken from promotional material from Alberta, from places visited and from information relating to animals of the region (see Figure 12.6). The inclusion of text historically on commemorative plates is quite common. In this

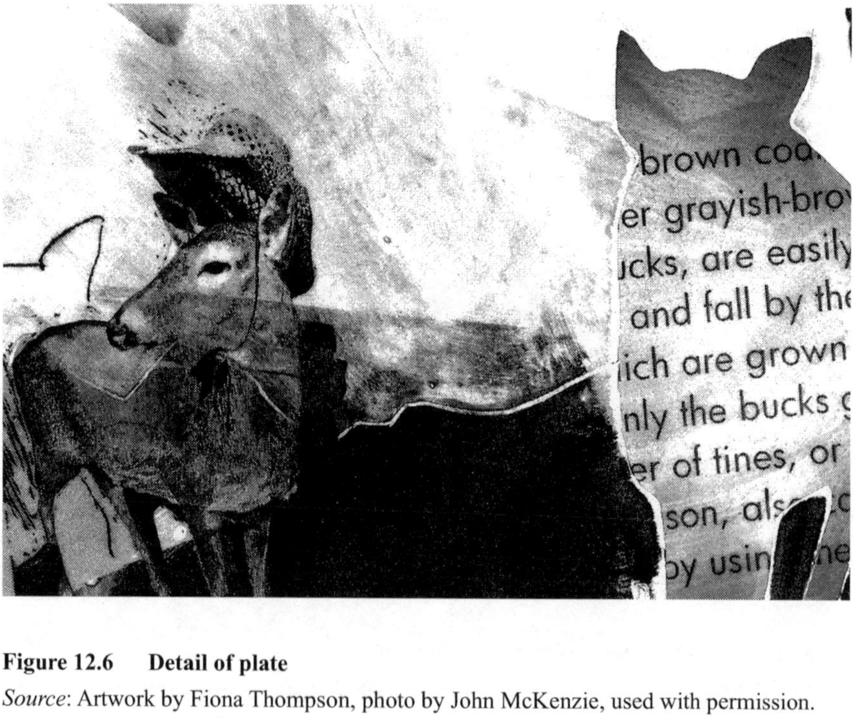

Figure 12.6 Detail of plate

Source: Artwork by Fiona Thompson, photo by John McKenzie, used with permission.

case it acts as another way of interpreting and communicating the subject matter. It can be read both visually and textually. The animals are simultaneously permanent (on the ceramic) yet appear fragile and ephemeral. Like their referents they are fixed permanently, frozen into a particular position or viewpoint. The plates in a sense commemorate the animals themselves and the experiences of travel in the area.

Conclusions

This chapter presented a dialogue between the act of travel, tourism theory, and the act of producing artwork. Moreover, the chapter offers an original and creative interpretation of taxidermic collections in Alberta, Canada through ceramic artworks. The context of the ceramic object as a vehicle for commemoration and narrative has been discussed giving further background to the production of artwork and its cultural location. The relevance of a focus on the particular plate format was emphasised. It also explored the engagement of the work with broader theories, in particular relating to hunting as a touristic activity in the region. As a collaborative project, the chapter discussed Fiona Thompson's responses to source material gathered from Alberta, as well as the academic material on tourism and hunting and the exploration of particular themes. The subsequent development of the artwork, which seeks to commemorate the animals themselves and the experiences of travel in Alberta, highlights the appropriateness of ceramic as a medium to interpret and communicate narrative, and goes some way to offer an insight into the artistic process.

References

Anderson, P. and Harrison, P. (eds). (2010). *Taking-Place: Non-Representational Theories and Geography*. London: Ashgate.

Baker, J. (1984). *Sunderland Pottery* (5th ed.). Sunderland: Thomas Reed Industrial Press.

Barber, L. (1980). *The Heyday of Natural History*. London: Butler & Tanner.

Baudrillard, J. (1994). The System of Collecting, in J. Elsner and R. Cardinal (eds), *The Cultures of Collecting* (pp. 7–24). London: Reaktion.

Beardsworth, A. and Bryman, A. (2001). The Wild Animal in Late Modernity: The Case of the Disneyization of Zoos. *Tourist Studies*, 1(1), 83–104.

Campbell, M. (2008). Communicating for Wildlife Management or Hunting Tourism: The Case of the Manitoba Spring Bear Hunt, in B. Lovelock (ed.), *Tourism and the Consumption of Wildlife* (pp. 213–226). London: Routledge.

Caroll, V. (2004). The Natural History of Visiting: Responses to Charles Waterton and Walton Hall. *Studies in the History and Philosophy of Biological and Biomedical Sciences*, 35, 31–64.

Dann, G. (1996). *The Language of Tourism*. Oxford: CABI.

Drakard, D. and Holdway, P. (2002). *Spode – Transfer Printed Ware 1784–1833*. Suffolk: Antique Collector's Club.

Foote, L. and Wenzel, G. (2008). Conservation Hunting Concept, Canada's Inuit and Polar Bear Hunting, in B. Lovelock (ed.), *Tourism and the Consumption of Wildlife* (pp. 188–212). London: Routledge.

Franklin, A. (1999). *Animals and Modern Culture*. London: Sage.

Godden, G. (1985). *Eighteenth Century English Porcelain*. London: Granada Publishing.

Godden, G. (2004). *Godden's Guide to English Blue and White Porcelain*. Suffolk: Antique Collector's Club.

Hannam, K. (2004). Tourism and Forest Management in India: The Role of the State in Limiting Tourism Development. *Tourism Geographies*, 6(3), 331–351.

Hannam, K. (2008). Shooting Tigers as Leisure in Colonial India, in B. Lovelock (ed.), *Tourism and the Consumption of Wildlife* (pp. 99–112). London: Routledge.

Hannam, K. and Knox, D. (2010). *Understanding Tourism: A Critical Introduction*. London: Sage.

Hare Duke, B. and Theophilus, J. (2011). British Ceramic Biennial Catalogue. Retrieved 20 April 2012, from <http://www.britishceramicsbiennial.com/>.

Lovelock, B. (2008). An Introduction to Consumptive Wildlife Tourism, in B. Lovelock (ed.), *Tourism and the Consumption of Wildlife* (pp. 3–30). London: Routledge.

Madden, D. (2011). *The Authentic Animal*. New York: St Martin's Press.

Norton, A. (1996). Experiencing Nature: The Reproduction of Environmental Discourse through Safari Tourism in East Africa. *Geoforum*, 27(3), 355–373.

Osborne, B. (2001). Landscapes, Memory, Monuments, and Commemoration: Putting Identity in Its Place [Draft paper available online]. Retrieved 15 February 2013, from <http://epe.lac-bac.gc.ca/100/200/300/citizenship_and_immigration_canada_metropolis/landscapes_memory-e/putinden.pdf>.

Pearce, S. (ed.) (1994). *Interpreting Objects and Collections*. London: Routledge.

Pearce, S. (1998). *Collecting in Contemporary Practice*. London: Sage.

Poliquin, R. (2010). Taxidermy. Retrieved 3 April 2012, from <http://www.ravishingbeasts.com>.

Rothschild, M. (2008). *Walter Rothschild: The Man, the Museum and the Menagerie*. London: Natural History Museum.

Russell, C. L. (1995). The Social Construction of Orangutans: An Ecotourist Experience. *Society and Animals*, 3(2), 151–170.

Scott, P. (2002). *Ceramics and Print London* (2nd ed.). London: A&C Black.

Scott, P. (2007). Propaganda, Politics and Porcelain. *Ceramics Monthly, February*, 34.

Scott, P. and Bennett, T. (1996). *Hot off the Press*. London: Bellew.

Sontag, S. (1977). *On Photography*. London: Penguin.

Thompson, F. (2010). *Representations of Tourism Expressed through Creative Ceramic Practice, with an emphasis on the Printed Photographic Image*. MPhil Thesis: University of Sunderland.

Thompson, F., Hannam, K. and Petrie, K. (2012). Producing Ceramic Art through Tourism Research. *Annals of Tourism Research*, 39(1), 336–360.

Whatmore, S. (2002). *Hybrid Geographies: Nature, Culture, Spaces*. London: Sage.

PART IV

Art, Place Image and Travel Motivation

Chapter 13

The Artist and the Tourist: Gauguin in Tahiti

Jean-François Staszak

Introduction

Tourism geographical imaginaries do not arise randomly, nor do they particularly rely on the touristic experience of a place itself. Rather, they exist prior to a tourist's visit, and it is when the tourist chooses his or her destination that this imaginary plays its greatest role. This image of a place and its people is formed from information learned at school, read in a book, a novel or a guide, or gleaned from images seen in newspapers, on the internet, or in museums, or viewed in films or on TV. Some of these texts and images are considered works of art; others are not, yet these too often refer, more or less deliberately, to artistic representations, by way of inspiration, imitation, quotation or parody. Consequently, Western (visual) culture, being notably indebted to artistic masterpieces, plays a significant role in the construction of tourists' imaginings of places of travel; these artistic masterpieces being the things that (tourists') dreams are made of.

With regard to tourism, the artist may acquire an even more seminal role. It has been said that some places were not considered as attractive, some landscapes not even gazed upon, before being represented in artistic masterpieces. As Roger asserts, 'No land ever primarily exists as landscape. The transformation of the former into the latter requires all the elaboration of art' (1997, p. 18).[1] For example, the Sainte-Victoire Mountain and more generally the landscapes of Provence in Southern France attracted little attention before being made famous by Paul Cézanne and Vincent Van Gogh at the end of the nineteenth century. Inventing the (taste for the) Mediterranean landscape, they opened the way for the development of tourism in Southern France.

(Great) artists are often perceived as playing such a role because they are seen to perceive and express their feelings about the world in a different way, ways which, after their artwork based on these has been produced, become popular. Conversely, tourists are often perceived as being less inventive than artists, and as following the paths of these great artists. However, I do not completely agree with this view, mainly because it is based on an implicit hierarchy or dichotomy between

1 Please note that all translations of quotes from French texts in this chapter are the author's own translations.

artists and tourists, which in some cases makes little sense, as far as the history of art and of tourism are concerned. Why should the artist's experience of place be necessarily genuine and authentic, and the tourist's fake and inauthentic? This idea is based on a common romantic and elitist ideology, shared by many tourists, which idealises the artists and despises the tourists. I suggest that the tourist's and the artist's geographical imaginary may actually have a lot in common.

In this chapter I claim that the distinction between the artist and the tourist is not always easy to draw and that the role played by some famous artists in the 'invention' of touristic landscapes may have been overestimated. Paul Gauguin is a much celebrated and revered artist, whose image and late paintings are identified with Tahiti (despite the fact that he moved from Tahiti to the Marquesas in 1901, his name nevertheless remains linked to the most famous island of French Polynesia). I suggest that considering Gauguin as a tourist may give a better understanding of his life and work, and of the development of tourism in French Polynesia. This rather iconoclastic hypothesis doesn't necessarily lead to questioning the artistic value of his paintings. I am not suggesting he was not a great artist (the value of his work, especially in Polynesia, is incontestable (Shackelford and Frèches-Thory, 2004)), but rather, that he may have been both a great artist and a tourist, and furthermore, that his Tahitian paintings are real masterpieces not *despite* him being a tourist but *because* he was one.

I will first consider what Gauguin knew or imagined about Tahiti before going there. I will then look at his life in Polynesia. Finally, I will conclude with some remarks about Gauguin and contemporary tourist imaginaries. However, before that I think it is fair to contextualise my own experience of, and relation to, Polynesia and Paul Gauguin, and to inform the reader about my position regarding the place and the person. My experience of French Polynesia, where I have visited four times (in 2003, 2005, 2006 and 2010), was firstly that of a male French geographer concerned with postcolonial and gender issues while researching Paul Gauguin or attending academic meetings in Papeete; secondly that of a male French tourist when I was naively visiting the islands or delightfully swimming in the lagoon; and thirdly that of a male French novice tour operator when I was lecturing about Paul Gauguin on the *Paul Gauguin* cruise ship. This encompassed guiding wealthy American tourists to Hiva Oa, from the ship to the site of the artist's grave and to the local museum dedicated to his life and work on the island, destinations that the tourists and myself were discovering at the same time. Indeed, during my short stay in Hiva Oa (and perhaps in many other circumstances), I fulfilled all of these roles simultaneously. This may explain how and why I came to question who is a tourist and who is not.

My work on Paul Gauguin and his geographical imaginary is based on the analysis of five different types of sources: his (abundant) private correspondence; his theoretical and literary writings; his artistic work; testimonies of his contemporaries; and the extensive literature from his biographers (especially Danielsson, 1988; Eisenman, 1997; and Sweetman, 1995). In addition, I also draw on my own book-length publications on Gauguin (Staszak, 2003, and 2006a). While continuing to admire the artist and to hold great affection for his work, I became more critical of Gauguin the person as I realised, at a relatively late stage and partly on the occasion of my first stay in Tahiti, that he was far from being the *Grand Homme* in Polynesia that he is taken for in France (see for instance Spitz, 2003).

Gauguin's Project and Geographical Imaginary

Gauguin was not the first European painter to sail to Tahiti. Several English and French artists, for example William Hodges (1744–1797) and Charles Girault (1851–1932), visited Polynesia before him. Hodges travelled with James Cook as the official painter for his expedition: his task was to provide the English public with images of these new lands (Quilley and Bonehill, 2004). Gauguin's project was quite different, indeed coming much later to Tahiti (1891), he was not an explorer in this sense. As such, to understand what he was looking for requires greater consideration of the place of Tahiti as it existed in the European imaginary from the 1770s to the 1880s.

The island was visited or 'discovered' by Samuel Wallis in 1768. The French explorer Louis Antoine de Bougainville arrived a year later, and it was he who made the island famous. Bougainville had read Rousseau and knew about the theory of the '*bon sauvage*'.[2] Tahiti for him was the living illustration of the idea that human beings, before becoming 'corrupted' by civilisation, enjoyed an easy and happy life. According to the first impressions of Bougainville, the Tahitians knew nothing about original sin and enjoyed guilt-free sexual freedom, an idea based on a tragic misunderstanding (Baré, 2002; Salmond, 2010; Tcherkézoff, 2004) but which nevertheless located the island at the very heart of the Western erotic imagination. On April 15 1768, he wrote in his log book:

> Nature has given Tahiti its laws. Tahitian people follow these rules in peace, and are maybe the happiest society on Earth. Politicians and Philosophers, you should come here and see accomplished what your imagination did not even dream of. As long as I live, I will celebrate the happy island of love. It is the true utopia. (Bougainville, 1771)

When Bougainville returned to France and wrote about Tahiti, it became a philosophical issue. Discussing Tahitian society from the Enlightment

2 Since the end of the eighteenth century, philosopher Jean-Jacques Rousseau (1712–1778) has been erroneously associated with the idealisation of the 'state of nature' and the '*bon sauvage*' ['*noble savage*'] myth, despite the fact that he did not use the term 'noble savage'. Indeed, Rousseau conceived the 'state of nature' as a hypothetical situation of human existence without society. He did not define this hypothetical situation as good or bad, draw parallels with the lives of the newly discovered 'savages', nor did he suggest that humankind should, or could, go back to the 'state of nature'.

perspective, Denis Diderot wrote the *Supplément au Voyage de Bougainville (Addendum to the Journey of Bougainville)* in 1772. Interestingly, the covers of several contemporary editions of his book are illustrated with Gauguin's paintings. Despite the anachronism, it makes some sense to illustrate Diderot with Gauguin, as the latter was much influenced by the philosopher.

The second important moment in the construction of Western, and especially French, imaginaries about Tahiti took place at the end of the nineteenth century. In 1880, Tahiti became a French colony, subsumed into the French Empire. The very same year, Julien Viaud (whose artistic pseudonym was Pierre Loti) published *Rarahu*, later reprinted as *Le Mariage de Loti [The Marriage of Loti]*. The novel tells the love story of the narrator, a French Navy officer, with a young vahine during his stop in Tahiti. The tremendous success of the novel culminated in it being one of the greatest best-sellers published in France. Because of the exoticism of the scenery and the eroticism of the narrative, Tahiti became a fashionable and exciting fantasy. For French and European readers of Bougainville, Diderot and Loti, Tahiti was the island of love and freedom, the home of the happy savages. Thus, when Gauguin sailed to Tahiti ten years later, in 1891, Tahiti was already firmly established as an attractive site in the French geographical imagination.

The notion of leaving Europe and forming an artist studio in the Tropics was suggested to Gauguin by Vincent Van Gogh during their common stay in Arles in 1888 (Druick and Zegers, 2001). In March of that year, three months before Gauguin's visit, Van Gogh wrote in a letter to his sister: 'I can imagine that a painter of today might make something like one finds described in the book by Pierre Loti, *Le Mariage de Loti*, where nature in Otaheite [Tahiti] is described' (Van Gogh, 1888, text in brackets added). In pointing to Tahiti, Van Gogh recognised the potential of capitalising on Loti's success. However, Van Gogh was not the first to suggest Tahitian paintings, for example in a popular magazine dedicated to travel and geography, which published three papers on Polynesia (1860, 1875, 1876), an explorer wrote:

> I would like to analyse the feelings that the sight of these scenes of picturesque beauty, nearly unique in the world, evoked in me during my stay [in Tahiti] … And the local people! How they animate this luscious spectacle! … What a picture for an artist! (Paihlès, 1876, p. 86)

Gauguin never acknowledged what his project (and his art) owed to Van Gogh, nor did he ever, at any stage, mention Loti as a source of inspiration. As an artist aspiring to radical originality, Gauguin could not admit that he was following Loti's path. As arguably occurs with many tourists, Gauguin was also a victim of the 'Armstrong syndrome' (Urbain, 1993), wanting to be the first in the place he visited.

If Gauguin's Tahitian artistic project was not entirely original, what about his geographical imaginary? Before sailing to Tahiti, Gauguin read many books where he gathered information about French Polynesia, which he then shared with the journalists to whom he explained his project and the friends and artists he had hoped (unsuccessfully) would join him. For instance, in August 1890 he wrote to Emile Schuffenecker:

> While on the opposite end of the earth, men and women meet their needs only after ceaseless labour, struggling with the rigours of cold and hunger, beset by misery and every kind of deprivation, Tahitian men and women, on the contrary, happy inhabitants of the unknown paradises of Oceania, know nothing but a sweet life. For them, to live is to sing and to love. (Merlhès, 1995, p. 48)

Gauguin's geographical imaginary of Tahiti seems quite stereotypical, finding obvious inspiration in Bougainville, Diderot and Loti. Interestingly, the sentence in his letter to E. Schuffenecker is not Gauguin's. It is an exact quotation of a book of colonial propaganda, published for the Universal exhibition in Paris in 1889 (Henrique, 1890). At the time, French society had not fully embraced colonisation, and the government endeavoured to convince French (mainly male) citizens to leave Europe and join the French empire (Blanchard and Lemaire, 2003).

Gauguin preconceived Tahiti as a lost paradise, where native people live innocently and happily, beholden only to natural laws, not only because he shared the geographical imaginary of his time, but also because he was a (possibly consenting) victim of colonial propaganda. The visual exoticisation and erotisation of Tahiti did not begin with Gauguin's paintings. To understand Tahiti's (sex) appeal and the importance of images in the process even before Gauguin, I suggest going back to Loti, who had much in common with Gauguin. In his autobiography, published in 1890, one year before Gauguin's departure to Tahiti, Loti recalls his readings as a young boy:

> [My brother] gave me as a present a big gilded book which was precisely *A Voyage to Polynesia* [most probably Delessert 1848], with many pictures, and it was the only book that I loved in my early childhood. I flipped though the book at once with immediate curiosity … I spent many hours of my break time to paint the images … With extreme care, I first colored the flowering branches, the groups of birds … As for the two *Young Tahitian women at the Ocean's Edge*, I made them white, oh white and rose-colored, like the sweetest dolls. (1890, p. 96 and 110)

It is worth noticing that these young Tahitian women are half-naked (see Figure 13.1) and that the book was published in 1848, which implies that erotic and exotic engravings of Tahitian vahines existed for more than 40 years before Gauguin sailed to Tahiti.

Figure 13.1 Young Tahitian women
Source: Delessert, 1848, p. 266.

One hundred years after the island's 'discovery', the European imaginary of Tahiti was rich and attractive (Bachimon, 1990; Boulay, 2005; *Kannibals* and *Vahinés*, 2001; Salmond, 2010), and Gauguin's fantasy about the island was nothing but a common (exotic and erotic) dream of the 1880s. But actually going to Tahiti in order to realise this dream nonetheless was exceptional. In spite of the early development of world-scale international tourism in the second half of the nineteenth century, and despite efforts of the French government to promote colonisation, Tahiti remained distant and isolated; indeed it took Gauguin 70 days to sail from Marseille to Tahiti in 1891. At the time, the island received very few visitors. If Gauguin came to Tahiti as a tourist, he was among the first. Gauguin's project was rather that of a coloniser or an orientalist (Said, 1979): he wanted to live and work in Tahiti, taking advantage of an easy life and finding inspiration in the local landscape and primitive arts. This dimension of his endeavour may be quite audacious and singular, but it was nevertheless based on geographical imaginaries of Tahiti which were already common and stereotypical, and similar indeed to those of would-be tourists.

In Tahiti, Gauguin would face disappointment similar to the disjuncture between imaginings and actual experiences that often occurs when tourists arrive at their destinations.

Gauguin's Work and Life in Tahiti

Gauguin never failed to project the authenticity of his experience in Tahiti, where he claimed to have lived as a native among the natives. But his assertions in this regard are questionable. Gauguin was one of the first artists to understand that, in order to sell his paintings, he had to sell his image. Successfully constructing his own myth, he worked hard to convince Western art critics, sellers and clients that he had become a real Tahitian. In fact, most of the time, he was living in or close to Papeete, the main Polynesian urban centre and the headquarters of the colonial administration. He found his friends and clients exclusively among the European community, although Gauguin would turn more native when in Hiva Oa (Marquesas Islands 1901–1903).

Arguably there exists little of Tahiti in Gauguin's work. For example, *Ia Orana Maria [Hail Mary]* (see Figure 13.2) is one of his first famous Tahitian paintings. The Polynesian title provides an exotic touch, which sounds quite artificial. Gauguin, particularly at that time, knew little of the Tahitian language and it is

Figure 13.2 ***Hail Mary*, Paul Gauguin, 1891**

Source: Reproduced with permission from The Metropolitan Museum of Art, New York.

likely that the title is taken from the missionary literature (missionaries going hand in hand with colonisation and serving as intermediaries between European travellers and native population). The exoticism of the scenery rests on (i) the landscape: the lagoon and the palm trees in the background, the local fruit in the foreground (two kinds of Tahitian bananas and breadfruit), the wooden dish filled with bananas and the boathouse and (ii) the models: their ethnicity, their pareo sarongs and the strange way Mary carries her child on her left shoulder.

The inspiration of this painting, I believe, has other geographical and artistic origins. The Angel and the delicate flowers before him show the influence of Italian Renaissance artists, like Botticelli. The two women in the middle are inspired by figures of dancers of the Javanese Borobudur temple most likely inspired by photographs of the temple which Gauguin bought when he was visiting Paris Universal Exhibition in 1889 (Druick and Zegers, 1991). Although it has been believed that the way Mary carries her son was something Gauguin may have observed in Tahiti, in fact, it is just a copy of a postcard he bought in Port Said, while his boat was making a stop at the Suez Canal, en route to Tahiti. And he was not the first painter using that body posture to exoticise his models (see for instance Léon Bonnat's *An Egyptian Peasant Woman and Her Child*, 1867, New York, Museum of Modern Art).

As soon as Gauguin arrived in Tahiti, he endeavoured to obtain items of primitive art, albeit it was almost impossible to find any in Tahiti. This was firstly because Tahiti had never been an important centre for the production of primitive art, unlike the Marquesas Islands. Secondly, most of the items had been destroyed or hidden when the local people were evangelised at the beginning of the nineteenth century. Gauguin could not even find anyone to tell him stories of the Polynesian gods and heroes. So there was no *tiki*, no ritual, and no story he could insert in his paintings. That is probably the reason his first paintings in Tahiti do not owe much to Polynesian arts and culture. More would come later, but it never came from his own experience of Tahitian culture: he found all the information in Western ethnographic books and during his visit to the Auckland Museum in New Zealand, in 1895. Considering where Gauguin found his information, it could be concluded that he was definitely more a tourist than an ethnographer. The kind of scenery, from an artistic point of view, that attracted Gauguin during his stay in Polynesia, is quite similar to the images on the postcards sold in Tahiti. The style is of course quite different, but the picturesque and exotic scenes he chose to depict are similar (Tréhin, 2003). Even in 1902, when Gauguin lived in the Marquesas Islands, he still may have found inspiration in postcards (see Figures 13.3 and 13.4). The similarities between Gauguin's paintings and postcards do not disqualify his work, but show the degree to which his imaginary and his view of Tahiti paralleled those of other travellers and photographers.

However, this was not something Gauguin wished to be known for. Explaining his inspiration for *Pape Moe [Mysterious Water]*; (1893 private collection), Gauguin wrote in his (much contrived) autobiography (Gauguin, 2001) that he went to the mountain in Tahiti, where he saw a young woman drinking water from a cascade in a sacred Polynesian site, mysterious scenery he tried to figure in this painting. In fact, Gauguin's painting was conspicuously inspired by a photograph

Figure 13.3 Two Women, Paul Gauguin, c. 1901 or 1902

Source: Reproduced with permission from The Metropolitan Museum of Art, New York.

Figure 13.4 Postcard from the beginning of the twentieth century

Source: Photograph by Henri Lemasson, 1897.

shot in 1888 by Charles Spitz, which he bought in a Tahiti curio shop. This doesn't question the quality of Gauguin's work, but rather the way he wished it to be perceived. Gauguin's refusal to admit that he drew his inspiration from a postcard and his apparent lies about his serendipitous encounter takes us back to questions of authenticity. Gauguin claimed to have direct experience and knowledge of local life and people, seeking to elevate the originality of his work and distinguish himself from the local colonisers or tourists and, correspondingly, between his art and postcards. But Gauguin's claim for a rare and authentic experience of the 'real' local people fails to differentiate him from tourists: they look at exactly the same things. And when tourists return home and want to tell the story of their journey and the places visited, they too may rearrange the truth to suit expectations and to make their experience more authentic and adventurous, just as Gauguin did. Denying one's status as a tourist can be seen as a typical tourist attitude.

Given the length of time Gauguin spent in Tahiti (eight years) he could not be classified as a tourist and he was certainly much more a coloniser than a native. However, his interest in Tahitian culture and landscape was similar to that of tourists. One could argue that this assertion is anachronistic, considering there were almost no tourists at that time in Tahiti. Still, it was not because tourism did not yet exist, or the place was not attractive, but rather because it was out of reach for most people. The development of tourism in French Polynesia had to wait for the opening of a regular steamship connection between California and Tahiti in 1924 and the Tahiti international airport in 1961. A few wealthy travellers did nevertheless make stops in Tahiti, buying the same postcards which inspired Gauguin.

If Gauguin was in some ways a tourist in Tahiti, what kind of tourist was he? He was surely anticipating ethnic (I'll come back to this later) tourism, but sex tourism as well. Loti confessed the importance of the erotic appeal of Tahitian vahines to his imagination of Polynesia, but Gauguin wrote nothing about it before sailing to Tahiti. He mentioned his sexual expectations when he considered other destinations (Madagascar) and aimed to convince in June 1890 his friend, Emile Bernard, to join him in his project: 'women down there are, so to speak, inevitable' (Gauguin, 1992, p. 192). In Tahiti, Gauguin was a regular client for local sex workers, and in the context of his times, Gauguin was perhaps the first sex tourist in Tahiti.

Gauguin and Contemporary Touristic Imaginaries

What does the recent development of tourism in Tahiti owe, exactly, to Gauguin? Gauguin's paintings present a beautiful image of Tahiti. Being well known and widely appreciated, they are free advertising for tourism in Polynesia. Gauguin's work and the myths they construct are utilised to promote Tahiti as a tourist destination. Tourist guides and tourism brochures often refer to Gauguin. Many tourists, if not most of them, have seen Gauguin's paintings. Gauguin's paintings remain central to tourist imaginaries about Tahiti.

But this does not mean that Gauguin was essential for the touristic development of the island. Tahiti was famous and attractive before Gauguin. Gauguin's paintings provided another set of representations which join Bougainville's journal, Loti's novel and, later, Frank Lloyd's and Lewis Milestone's *Mutiny on the Bounty* (1935, 1962) in putting Tahiti on the map of Western imaginaries, locating a lost paradise and a major Western social utopia in Polynesia. These myths have a longer history, going back to Rousseau's supposed noble savage, the biblical Eden and the Greek Golden Age. Gauguin's work is surely revolutionary as far as art history is concerned, but he nonetheless painted a highly stereotyped and fanciful Tahiti. Contributing to the Western imaginary about Tahiti, Gauguin added his bright colours, he made it more vivid, but he did not invent or change it; he merely gave a new expression to an old dream. The painter, however, is so famous and his work so celebrated and reproduced that, for a tourist or anybody, it is nearly impossible to imagine or even to see Tahitian landscapes and people without reference to Gauguin's art. And this is not without problems or consequences:

Firstly, the landscapes or the vahines made famous by Gauguin's paintings can hardly be found in Tahiti. Most Tahitian women look quite different from Gauguin's models, and they surely are not half-naked. Tahitian society and landscape is different from the way they appear in Gauguin's paintings not only because Tahiti has changed since Gauguin, but also because Gauguin did not paint Tahitian reality but rather his own Tahitian fantasy, which is also the fantasy of most tourists. Correspondingly, Polynesian luxury resorts, which seek to meet visitors' expectations, do not encourage independent exploration beyond their grounds and are 'physically managed to replicate an imagined space' (Kahn, 2011, p. 131; Staszak, 2006b). Wealthy tourists, particularly, have few opportunities to confront the 'real' Tahiti, and little chance of having their imagination falsified.

Secondly, through my own experience and knowledge of Tahiti, I encountered Tahitian people who told me that they do not share the Western fascination with Gauguin and indeed there are those that question the representation of Tahiti in his work. I also gained the impression that his work was perceived by some Tahitians to present a false image of Tahiti. I detected some resentment about the fact that he became rich and famous as a result. Indeed, a predominant feeling among the Tahitians I met was also that they did not want the tourists to see their society and island through the lens, or brush, of a man who was, according to some, nothing but a colonial painter.

These two issues relate to major discrepancies between Gauguin, his work and Tahiti. A third one deals with a discrepancy between Gauguin's imaginary and practices and today's tourists. Gauguin almost never represented Tahiti's lagoon and its blue waters, and indeed only sparingly used blue in his paintings (Vacher, 2013). At the time, no one was particularly interested in the lagoon or the possibility of bathing in it. Gauguin and the local people were bathing in Tahitian lakes and rivers. Today, the lagoon is the main attraction. Most of the postcards one now finds in Tahiti show the pristine blue lagoon and have little resemblance to Gauguin's paintings (or any of the turn-of-the-century postcards). If today's postcards still show half-naked vahines, the background is different. The image of the lagoon, an essential element of the current Western imaginary of Tahiti, is too recent to owe anything to Gauguin. Tahitian people, I believe, might at times feel ignored, even disparaged by the tourists who sail on the Paul Gauguin cruise ship, buy Paul Gauguin postcards and mugs, and, when they aren't bathing in the lagoon, visit the Paul Gauguin's museum and run to Paul Gauguin's grave, without paying much attention to local culture and people. If Gauguin's paintings are a billboard for Tahiti, they also stand as a screen and obstacle between the tourists and the actual people of Tahiti. But does Tahiti even exist? According to a website promoting Tahiti to North American tourists in 2002, the island is 'a destination so exotic, it exists only in one's dream' (www.GoToTahiti.com, 2002 in Kahn, 2003, p. 308)

When asked in 2005 the reason why they chose Tahiti as a destination, 20 per cent of the tourists said they were following a dream (KPMG THL, 2005). One could say they are following Gauguin's dream, not because Gauguin invented Tahiti but because they share the same imaginary, which was built in Western Europe at the end of the eighteenth century. And the most recent slogan for the strategic development of Tahitian tourism, *Polynesia, the last Garden of Eden*, will not help change it (Tahiti Tourism Authority, 2011).

The genealogy of contemporary tourist iconography could be traced to the engravings following the 'discovery' of the island, continuing on to Gauguin's paintings, turn-of-the-century postcards, chocolate wrappers, *National Geographic* illustrations and advertising photographs. These images show a never-ending fascination for the welcoming half-naked vahine (see Figures 13.5 and 13.6). Contemporary publishers of postcards perpetuate this image, their best-selling pictures being Gauguin's paintings and photographs of attractive half-naked women, who pass for vahines merely by having a crown of leaves put on their head. Defending their sale of erotic images that many may consider offensive, they explain that they just 'create images that [they] think people want to buy' (Kahn, 2011, p. 80–82), exonerating themselves and placing the blame on the tourist. They nevertheless play an active role in the exoticisation and eroticisation of Tahiti, like that of Gauguin. Their impact is the same, and their motivations are similar. Both take part in (re)producing a Western geographical imaginary and (re) creating the landscape and the people that we then eventually seek in Tahiti.

Conclusion

Gauguin, in his life and work, blurred the line between the tourist and the artist. To assume that he was essential to the touristic development of the island would be a mistake. Still, Gauguin remains a major attraction for tourists in Tahiti. Not that there is much of his work or many traces of his presence in the island: two (rather disappointing) museums are dedicated to his life and work, one in Tahiti and one in Hiva Oa. Tourists buy all kinds of items reproducing his paintings such as books, postcards, posters, stamps, magnets, mouse pads, bags, lighters, mugs,

Figure 13.5 Two Tahitian Women, Paul Gauguin, 1899

Source: Reproduced with permission from The Metropolitan Museum of Art, New York.

Figure 13.6 Advertisement from the 2010–2011 Air Tahiti Nui and Tahiti Tourisme joint advertising campaign

Source: Image source and copyright: tim-mckenna.com, reproduced with permission.

T-shirts, soaps, perfumes and chocolate boxes. Art connoisseurs and historians may condemn this vile merchandising, but there is an ironic logic regarding Gauguin's geographical imagination and expectations, in that he was not that different from contemporary tourists.

Exoticism did not rely on the (frightening) confrontation with the Other. The Other had to be transformed into a picturesque commodity in order to become the object of the exotic spectacle, which confirms the visitors' geographical fantasies and provides them with the reassuring experience of domesticated people and landscapes and the conspicuous evidence of their own superiority (Staszak, 2012). Colonisation and liberal globalisation played a major role in the domestication and commodification of the Other. In both narratives and images artists took part in rendering the Other picturesque and this in turn drew the attention of many tourists. In this regard, Gauguin responsibility cannot be denied. However, I believe that even without him, Tahiti would nevertheless have become a well-known international destination, although tourists might have come with other images in mind, and perhaps in fewer numbers.

Gauguin's major impact on tourism is indeed less direct, broader, deeper and more diffuse than that. His primitivism opened the way to the Cubists' and Surrealists' fascination for the *art nègre*, and to the (ambiguous) appreciation of 'primitive' cultures and people by Westerners. Gauguin was the first to affirm that their artefacts deserved to be kept in Art museums, not just ethnographic or natural history museums. His interest in the culture of 'primitive' people and his willingness to share their life not only changed art history but also laid the foundations for a new kind of tourism which would later be called ethnic tourism. Gauguin's influence is perhaps more discrete in Polynesia, where the lagoon remains a quasi-monopolistic attraction, than in other places (deep forests, sand deserts, high mountains and other tropical islands) where tourists, unsatisfied with modern Western civilisation, come to meet 'primitive people', re-performing the first encounter and hoping for spiritual rejuvenation, just as Gauguin did more than a hundred years ago.

Acknowledgements

I would like to thank Daniel Hoffman for his help in revising my English and for his critical reading of an early version of this chapter.

References

Bachimon, P. (1990). *Tahiti Entre Mythes et Réalités*. Paris: CTHS.

Baré, J. (2002). *Le Malentendu Pacifique. Des Premières Rencontres Entre Polynésiens et Anglais et Ce Qui s'Ensuivit Avec Les Français Jusqu'à Nos Jours*. Paris: Editions des archives contemporaines.

Blanchard, P. and Lemaire, S. (eds) (2003). *Culture Coloniale. La France Conquise par son Empire 1871–1931*. France: Autrement.

Bougainville, A. d. (1968 [1771]). Journal de Bord. *Journal de la Société des Océanistes*, 24, 11–34.

Boulay, S. (2005). *Hula Hula, Pilou Pilou, Cannibales et Vahinés*. Paris: Éditions du Chêne.

Danielsson, B. (1988 [1964]). *Gauguin à Tahiti et Aux Iles Marquises*. Paris: Presses Pocket.

Delessert, E. (1848). *Voyages dans les Deux Océans Atlantique et Pacifique*. Paris: A. Franck.

Diderot, D. (2002 [1796]). *Supplément au Voyage de Bougainville*. Paris: Folio.

Druick, D. and Zegers, P. (1991). Le Kampong et la pagode: Gauguin à l'exposition universelle de 1889. *Gauguin. Actes du colloque Gauguin. Musée d'Orsay* École du Louvre et Musée d'Orsay, Paris, 11–13 janiver 1989: La Documentation Française.

Druick, D. and Zegers, P. (eds) (2001). *Van Gogh and Gauguin: The Studio of the South*. New York: Thames and Hudson.

Eisenman, S. (1997). *Gauguin's Skirt*. New York: Thames and Hudson.

Gauguin, P. (1992 [1964]). *Lettres à sa femme et à ses amis, recueillies, annotées et préfacées par Maurice Malingue*. Paris: Grasset.

Gauguin, P. (2001). *Noa Noa*. Papeete: Avant et Après.

Henrique, L. (ed.) (1890). *Les Colonies françaises, 4ème vol., Colonies et protectorats de l'Océan Pacifique*. France: Quantin.

Kahn, M. (2003). Tahiti: The Ripples of a Myth on the Shores of the Imagination. *History and Anthropology*, 14(4), 307–326.

Kahn, M. (2011). *Tahiti Beyond the Postcard: Power, Place and Everyday life*. Seattle: University of Washington Press.

Kannibals & Vahinés. Imagerie des Mers du Sud. 23 octobre 2001–18 février 2002, Catalogue de l'Exposition. Paris: MAAO/RMN.

KPMG THL. (2005). Étude de Perception et d'image de la Destination de Tahiti, Mandatée par le GIE Tahiti.

Loti, P. (1991 [1880]). *Le Mariage de Loti*. Paris: Garnier Flammarion.

Loti, P. (1890). *Le Roman d'un Enfant*. Paris: Calmann-Lévy.

Merlhès, V. (1995). *De Bretagne en Polynésie. Paul Gauguin. Pages inédites*. Papeete: Avant et Après.

Paihlès, M. A. (1876). Souvenirs du Pacifique. L'archipel Tahiti, *Tour du Monde. Nouveau Journal des Voyageurs*, 1er semestre. Paris: Hachette.

Quilley, G. and Bonehill, J. (eds) (2004). *William Hodges 1744–1797: The Art of Exploration*. Greenwich: Yale University Press.

Roger, A. (1997). *Court Traité du Paysage*. Paris: Gallimard.

Said, E. (1979). *Orientalism*. New York: Vintage Books.

Salmond, A. (2010). *Aphrodite's Island: The European Discovery of Tahiti*. Berkeley: University of California Press.

Shackelford, G. T. M. and Frèches-Thory C. (eds) (2004). *Gauguin Tahiti: The Studio of the South Seas*. London: Thames & Hudson.

Spitz, C. (2003). *Où en Sommes-nous Cent ans Après la Question Posée par Gauguin D'où venons-nous? Que Sommes-nous? Où Allons-nous*. Paper presented at the Paul Gauguin. Héritage et confrontations. Actes du colloque des 6, 7 et 8 mars 2003 à l'Université de la Polynésie française, Papeete.

Staszak, J.-F. (2006a). *Gauguin Voyageur*. Paris: Solar/Géo.

Staszak, J.-F. (2003). *Géographies de Gauguin*. Paris: Bréal.

Staszak, J.-F. (2012). La Construcción del Imaginario Occidental del 'allá' y la Fabricación de las 'Exótica': El Caso de los Koi Moko Maoris, in D. Herniaux and A. Lidon (eds), *Geografia de los Imaginerarios* (pp. 179–210). Barcelone/Mexico: Anthropos/Universidad Autónoma Metropolitana Iztapalapa.

Staszak, J.-F. (2006b). Voyage et Circulation des Images: du Tahiti de Loti et Gauguin à Celui des Voyagistes. *Sociétés et Représentations*, 21, 78–99.

Sweetman, D. (1995). *Paul Gauguin: A Complete Life*. London: Hodder & Stoughton.

Tahiti Tourism Authority (2011). Stratégie de Développement Touristique <http://www.servicedutourisme.gov.pf/spip.php?article128>, accessed 02.05.2013).

Tcherkézoff, S. (2004). *Tahiti-1768. Jeunes Filles en Pleurs*. Tahiti: Au vent des îles.

Tréhin, J.-Y. (2003). *Gauguin, Tahiti et la Photographie*. Tahiti: Musée de Tahiti et des Îles.

Urbain, J.-D. (1993). *L'Idiot du Voyage: Histoires de Touristes*. Paris: Payot.

Vacher, L. (in press – 2013). The Marvel of Tropical Waters: The Invention of an Imaginary at the Pace of Technological Advances, in M. Gravari-Barbas and N. Graburn (eds), *Tourism Imaginaries: Place, Practice, Media*. Farnham: Ashgate.

Van Gogh. (1888). Letter 590, from Vincent to Willemien Van Gogh written on or about 30 March 1888. Retrieved 20 May 2012, from <http://vangoghletters.org/vg/letters/let590/letter.html>.

Chapter 14

Image, Identity and Tourism: Public Art in Cultural Quarters

John McCarthy

Introduction

Cities throughout the world are increasingly encouraging concentration of culture-related uses within specific areas, often labelled as 'cultural quarters' or 'cultural districts' in order to achieve use synergies and broadly-based regeneration outcomes. These outcomes may be linked to tourism since many cities are seeking to attract visitors specifically to such areas by promotion and signposting or area branding. This is in turn linked to wider city branding via the (re-)creation of city 'quarters', which may bring economic benefits (linked to visitor spending), environmental and social benefits. One mechanism often applied in such quarters is public art, since this has often been seen as a reliable method of attracting visitors (for instance by linking such art pieces to designated 'trails'), promoting place image (via city branding) and enhancing local identity/ies (McCarthy, 1998, 2006a, 2006b).

However, such activities give rise to a number of questions, including: are aims for promoting place image and local identity contradictory? How effective is public art in this process? These questions are explored in this chapter via case studies of public art schemes in two UK 'cultural quarters': Manchester's 'Northern Quarter', and Belfast's 'Cathedral Quarter'. This chapter is arranged as follows: first, 'culture-led' approaches to regeneration and issues of the use of public art linked to identity and image are explored; second, the case studies are set out; third, these are discussed and compared; and finally, broad conclusions and lessons are suggested.

'Culture-led' Regeneration Approaches, Public Art, Identity and Image

'Culture-led' approaches to regeneration, involving culture-related uses to achieve broader regeneration outcomes, can potentially lead to economic innovation and diversification (Williams, 1997; Mommaas, 2004; Scott, 2004), image enhancement (Ashworth and Voogt, 1990; Miles, 2005b) and social cohesion (Matarasso, 1997; Department of Culture, Media & Sport, 1999; Scottish Executive, 2000). These may involve cultural uses such as art galleries as well as linked uses such as restaurants, within mixed-use areas. 'Cultural quarters' may be designated (Wansborough and Mageean, 2000), often as part of an array of distinctive urban 'quarters' (Bell and Jayne, 2004). These areas may be defined in spatial plans which often seek to bring about further concentration of such uses in order to achieve use complementarities, synergies, agglomeration economies and minimisation of amenity loss (Evans, 2003; Richards, 2001a, 2001b). However, it may be argued that such quarters need to link more effectively to local identity (Montgomery, 1995, 2003; McCarthy, 2006a), since enhancing place or local identity can assist with aims for attracting tourists and visitors, and it can also assist with place branding priorities, though this can also arguably lead to loss of distinctiveness which conflicts with visitors' and tourists' desires for authenticity.

As part of such broad approaches, public art, or site specific art in the public domain, can be used to achieve broad regeneration initiatives (Hall and Smith, 2005; Miles, 1997). For instance, it can assist in promoting a city's image, partly to attract visitors and tourists (Ashworth and Voogt, 1990; Avraham, 2004). It can also improve city legibility (Porch, 2000); reflect reaction against the commodification of art (Miles, 1997); promote civic identity, community building and social network development (Hall and Robertson, 2001); promote cultural diversity (Hall and Smith, 2005); and improve public safety (Hall and Robertson, 2001). In this way, it can arguably transform areas fundamentally (Wansborough and Mageean, 2000).

Public art may also have specific benefits in terms of economic benefits arising from tourism and visitor numbers. In this context, such art would seem to link in particular to cultural tourism (Hall and Robertson, 2001; Sharp et al., 2005), which has been described as 'Tourism motivated wholly, or in part by interest in the historical, artistic or lifestyle/heritage offerings of a tourism destination' (Briggs, 2005, p. 3). This type of tourism has emerged to meet the demands of a segment of visitors who seek a particular type of visitor experience and this can link in particular to the visual and performing arts, heritage attractions, museums and galleries, artists' studios, cultural events as well as public art itself.

More generally in this context, public art may be used to enhance 'destination image' relating to a wide range of tourism and visitor profiles. For instance, in 2008 it was estimated that the 'Waterfalls' public art exhibit on the East River in New York, by Olafur Eliasson, generated $69 million in economic activity, exceeding earlier expectations of around $55 million. It was also estimated that 79,000 people were attracted to visit or extend their stay in the city as a result (Brown, 2008). This raises issues of the potential importance of scale, innovation and spectacle for public art, since, to provide clear benefits in terms of visitor attraction, public art arguably needs to provide an attraction in its own right. While an example such as Antony Gormley's 'Angel of the North' in Gateshead may arguably link to both historical identity and to a future inspirational vision, in other cases inspiration or innovation may be prioritised over local context so as to maximise increased visitor numbers. Such factors (linked to city image and promotion as considered below) may be critical to achieving (or exceeding) expectations in relation to visitor numbers and/or associated economic impacts, but they may conflict with other aims such as for reflection of local identity and context.

Moreover, many claims for public art may be contested. For instance, Selwood (1995) indicates that public art may fail to meet community needs, and Miles (2005a) highlights the ambiguity in some justifications of public spending on public art. Public art may also arguably be mechanistic and contrived, and may have the effect of de-politicising regeneration practice (Hamilton, Forsyth, and de Iongh, 2001). It may also undermine the traditional role of the artist as critic of the status quo (Miles, 2005a), as reflected by those who suggest that public art should seek to encourage dissent (Sharp et al., 2005). Indeed, it may also be argued that public art that is linked to major projects can contribute to the endorsement of developments which are environmentally damaging and socially exclusive (Hall and Robertson, 2001).

Public art is bound up with place identity, or shared characteristics (Miles, 2005b), allowing places to be endowed with meaning (Neill, 2004). While 'image' relates to people's impressions of a city, 'identity' relates to a city's history and circumstances, so image creation may underpin identity formation. But local identities evolve over time, particularly in marginalised spaces (Miles, 1997; Julier, 2005), and they may be multiple and possibly conflictual, with weak identity linked to social exclusion (Griffiths, 1999).

Discourses of marketing and place branding attempt to show how cultural capital can be used to promote a city to bring economic benefits for instance from visitors. Nevertheless, there seems to be little clear empirical justification of this, particularly linked to public art, and public art's aesthetic benefits involve a degree of subjectivity in perception of outcomes. It would seem however that public art can contribute to strategies for strengthening local identity/ies; as Miles suggests, 'investment in culture … can actually serve to revitalise the identities of the people of a city … [and] … offer a permanent legacy' (2005b, p. 921, text in brackets added). This is linked to city 'branding' strategies (Cameron and Coaffee, 2005; Ashworth and Voogt, 1990; Evans, 2003, 2004), which can reinforce local identity (Evans, 2003).

Overall however, the benefits of public art are arguably often assumed rather than demonstrated, with little objective monitoring of their effects; moreover, new 'brand identities' involve the danger of 'brand decay' (Julier, 2005), and place imaging strategies may conflict with identities where such strategies are formulaic, with serial replication possibly resulting in erosion of distinctiveness and authenticity (Miles, 2005b; Julier, 2005). Furthermore, the commodification of public art may be linked to gentrification (Zukin, 1995; Cameron and Coaffee, 2005). Consequently, some argue that public art should reflect the local context (Breen and Rigby, 1996; Porch, 2000), for instance by reflecting local histories (Miles, 2005b; Evans, 2004). Indeed, Hall and Robertson highlight the argument that such art:

> … is able both to articulate and strengthen links between communities and places … Such advocacy draws on the equation of homogenizing urban development or

> modernist planning and the erosion of senses of place. Public art is cited with the ability to replace a quality that has vanished from a place or has been ignored; this is achieved, it is argued, through communication of events or aspect of local history. (Hall and Robertson, 2001, p. 12)

Conversely, others suggest the need for public art to reflect modernity (Julier, 2005; Montgomery, 1995), since historical associations may simply reflect nostalgia (Sharp et al., 2005). In addition, while some argue for flexibility for (public) artists, this may lead to lack of community engagement, and institutional sponsors may require guidance for public artists to ensure diversity of provision (London Borough of Croydon, 2005; Bristol City Council, 2000); artworks at different scales (Sharp et al., 2005); and improved legibility via new landmarks, signposts or gateways (Miles, 2005a).

This chapter highlights these issues via research conducted by the author in two case study cities, selected as public art initiatives within a 'cultural quarter', with contrasting elements in each case. Both used documentary analyses and structured interviews with key public officials. The Belfast case is contained within a formally-designated cultural quarter (in a spatial plan), while the Manchester case occurs within an informal cultural quarter (acknowledged in city promotion documents). Moreover, while the Belfast case involves a high degree of control and guidance, that of Manchester illustrates a high degree of flexibility. The research explored the extent to which the aim of public art was to promote image externally (outside the city), or to reflect image internally (within the city), or both, and to what extent such aims could be integrated. The cases are considered below.

Manchester's Northern Quarter

Manchester's Northern Quarter comprises 56 acres to the north of the city centre. This area was historically the centre of the city's wholesale fashion industry. It emerged as an informal cultural quarter in the early 1990s as a result of the success of Affleck's Palace, a retail emporium, the Dry Bar and several live music venues (Wansborough and Mageean, 2000). Property was cheap compared with the city centre, and environmental improvements and public art were funded by the City Council, the European Regional Development Fund, and the National Heritage Lottery Fund (personal interview with City Council planning officer). The City Council wished to support creative uses together with housing and entertainment uses, and an important objective was to bring vacant floorspace back into productive use. Partly as a result, by 2005 the area had become a focus of creative industries including associated back-up services (personal interview with City Council planning officer).

The Northern Quarter Public Art Scheme was preceded in 1996 by an environmental improvement scheme on Tib Street in 1996. The Public Art Scheme aimed to develop artworks reflecting the area's unique cultural identity, and to contribute to employment, education and quality of life aims. A 'long list' of 50 artists led to 20 expressions of interest. Phase One included decorative art and applied art, and Phase Two included site markers and iconic gateway projects. The Tib Street Corridor was the core of the scheme, comprising many pieces integrated with pavements and walls, including street signs, poetry in paving, tiling work, and a neon light tower. An 'art trail' was also intended in part as a tourism attraction. The governing concept for the scheme was 'warp and weft', relating to the textiles trade in terms of the weave of cloth, the warp representing Tib Street and the weft representing cross-cutting streets. However, some artists suggested the art trail involved 'disneyfication' of the area, though an evaluation of the trail highlighted its potential to act as a visitor attraction (Nowakowski and Walker, 1999).

The scheme used local history to reflect local identity, and the City Council provided a historical portfolio of the area, highlighting textiles, boot and shoe manufacture, hat and shoe making, and selling of pets. These were reflected in several artworks, for instance in ceramic plaques and murals, iconic motifs of birds, and a sculpture made of recycled industrial parts, although the largest artwork, a 90 foot high neon light tower by Peter Freeman, linked to the recent growth of evening uses. The cost of artworks ranged from £500 (for inset tiles) to £30,000 (for the light tower).

The scheme was co-ordinated by lead artist Liam Curtin, and initially managed by a partnership of the City Council and the Northern Quarter Association (a self-organised group of 130 members including local traders and residents) (Wansborough and Mageean, 2000). Funding was initially provided by the National Lottery Heritage Fund (£200,000), the European Regional Development Fund and the City Council, and local development company Urban Splash (personal interview with planning officer). Decisions on artworks were made by the Northern Quarter Management Committee, and a Steering Group. Commissioning of public art was flexible, with an 'open' brief allowing the artists to choose their own sites for artworks in a different way from other public art schemes in the city (Nowakowski and Walker, 1999).

While the flexible strategy allowed an organic approach, the scheme might have benefited from a more structured framework as well as more detailed guidance (Wansborough and Mageean, 2000), and some artists indicated confusion over expectations. Moreover, an evaluation of the scheme, commissioned to consultants by the City Council, recommended a clearer strategic project plan with a tighter commissioning process (Nowakowski and Walker, 1999). In terms of consultation, the procedure involved local residents and businesses, though linkage of consultation results to artworks was confusing to some (Nowakowski and Walker, 1999), though the mix of works was popular and was seen to raise the area's profile. The evaluation indicates strong local identification with the scheme, but also the danger of an over-reliance on a 'localist' approach, and it recommends more signature artworks and major commissions to raise the area's profile. It also acknowledges the potential conflict between aims for local identity and external image, and some artists suggested that use of historical references was prioritised over artistic quality, innovation or impact (Nowakowski and Walker, 1999).

Belfast's Cathedral Quarter

Belfast's Cathedral Quarter, around St Anne's Cathedral, is one of the oldest districts in the city, though decline in the 1980s led to the area becoming attractive to artists and community arts organisations. 'Cultural quarter' designation was proposed by the Laganside Corporation, a public regeneration body used public investment to secure private development capital, using powers to acquire land and carry out works. The Corporation existed from 1989 until 2007, when responsibilities were transferred to the devolved government's Department for Social Development. Cultural quarter designation was supported by local people, and the aim was to create a mixed-use area including for instance artists' studios, art galleries and a community arts forum (Laganside Corporation, 2003). The area was called the Cathedral Quarter, and identified in the City Council's Belfast Metropolitan Area Plan (Belfast City Council, 2004).

The Cathedral Quarter Strategy sets out regeneration aims (Laganside Corporation, 2000), including to encourage more arts groups in the area. The Corporation's Development Framework for the area identifies key development projects, including mixed uses with active ground floor uses, car parking facilities, apartments, offices, retail and leisure uses, additional managed workspaces, and a theatre complex (Laganside Corporation, 2001).

Aims related to public art included setting out a clear vision and the use of clear benchmarks and quality standards. Unlike the Manchester case, public art in the Cathedral Quarter did not comprise a separate scheme, but part of the overall art provision within the Laganside area. In this context, the Laganside Public Art Strategy (Laganside Corporation, 2001) indicates the need for public art that contributes to local identity and sense of place, and is recognisable, and detailed aims are contained in a public art Brief (Laganside Corporation, 2000).

The Public Art Strategy sets out criteria for selection, including quality of design, imagery, materials, colour, texture, context, function and use of site. It also indicates that public art should reflect local history, and include community artworks. But an emphasis on regeneration of development sites in the Corporation's Corporate Plan is reflected in the Public Art Strategy's terms of reference (Laganside Corporation, 2002), which also shows the need to work in broad partnership with landowners and developers. Nevertheless, the Corporation also aims to use public art to reflect a diverse community in terms of race, religion and disability, assisted by a robust artist brief and community input (Laganside Corporation, 2002).

Historical associations, as in the previous case, were evident in many artworks, as required by the Cathedral Quarter Public Art Brief (Laganside Corporation, 2000). This is reflected in cast iron sculptures highlighting the theme of commerce, and a ceramic mural showing a historical map of the area; other artworks include a small performance space, and steel railings reflecting the literary theme of Writer's Square. As in Manchester, the artworks form part of a trail, with two other public art trails in the Corporation's area, linked to the Corporation's aim to attract visitors (personal interview, Laganside Corporation officer).

The public artworks were managed by the Laganside Corporation, and the Arts Council for Northern Ireland and the Laganside Corporation each funded 50 per cent of works on average. The public art brief was widely distributed (personal interview, Laganside Corporation officer), and the brief indicates that artworks should seek to reflect historical elements but also express contemporary art practice, bearing in mind that the area is a Conservation Area with a distinctive historical streetscape. Works were also required to involve low or minimal maintenance, and artists needed to work closely with the design team (Laganside Corporation, 2000). Three commissions were initially offered, with requirements including use of historic features such as printer's blocks associated with the printing trade (Laganside Corporation, 2000).

The design team included Corporation staff, a consulting engineer, a landscape architect, civil, mechanical and electrical engineering advisors, a quantity surveyor and the statutory authorities. The selection panel included the Sculptural Society for Ireland, the Arts Association for Ireland, the Royal Ulster Architectural Society and Belfast City Council, and it was chaired by the Laganside Corporation. It applied criteria such as quality and professionalism, and many selected artists were local (personal interview, Laganside Corporation officer).

Local communities were represented on the Advisory Panel, and a consultation document prior to the Public Art Strategy was sent to 50 organisations for comment. Attitude surveys showed that public art in the area was viewed positively, and the artworks would seem to have contributed to wider regeneration, as indicated for instance by an increase in the number of arts organisations in the area from 20, before designation, to 50 in 2005 (personal interview, Laganside Corporation officer).

Public Art, Place Image and Identity

These two cases demonstrate aspects of the linkage between public art and local identities, with implications for tourism attraction. Three issues may be identified, namely aims for public art, context sensitivity, and the mechanisms applied. In terms of aims, the Northern Quarter scheme's aims were wide, reflecting elements such as vibrancy and innovation, creativity and distinctiveness. Similarly, in Belfast the brief indicates the need for artworks to respect the context, as reflected in works such as the iron railings in Writers' Square. Hence in both cases public art was aimed at contributing to a distinctive character as well as broader regeneration aims.

While the Northern Quarter public art scheme's aims do not specifically include image enhancement, the area represents 'Manchesterness' via its design orientation (Julier, 2005), and the City Council identifies the Northern Quarter in promotional literature for visitors. Thus the public art scheme would seem to have involved a re-orientation of image, albeit with some tension with aims for local identity. In Belfast, there was an explicit attempt to enhance the area's image.

Hence image enhancement played an important part in both cases, but particularly so in Belfast. Both schemes also explicitly used 'art trails' to promote tourism (personal interview, Laganside Corporation officer; Nowakowski and Walker, 1999). While some Manchester artists feared cultural commodification, the cases do seem to have been generally mindful of the need for a degree of authenticity and artistic innovation.

Institutional sponsors of public art initiatives played a critical role here. In Manchester, public art was managed with close involvement of the City Council, while in Belfast the Laganside Corporation managed the process which was focused on physical development. In the latter case the primary aim related to public realm improvement and private funding for development, based on models such as Newcastle's Urban Development Corporation and Birmingham's Brindley Place (personal interview, Laganside Corporation officer). Nevertheless, both cases applied an approach which included development promotion. In Belfast, this was clear at the outset, since the city had suffered for a long period from a negative image, but in Manchester a promotional emphasis emerged from increased culture-related activities.

In terms of reflection of local identities, there was an emphasis in both cases on use of artworks to reflect local history, largely because both cases involved distinctive historical associations. This was contested to some extent in Manchester since it was seen to represent nostalgia, but in Belfast the cultural and religious divisions allowed the area's (shared) historical associations to become a focus for consensus. This suggests that reflection of authentic associations of the local context can be popularly supported even if local identities are fragmented, diverse or divergent. Moreover, use of historical associations may provide links to local heritage and may also provide educational benefits, as well as a focus for celebration and commemoration. This would seem to link also to aspects of cultural tourism in terms of the potential desire of cultural tourists for authenticity and reflection of local context.

In terms of procedures and mechanisms, the Manchester case used a loose, fluid and flexible process, with 'negotiated' briefs, and innovative participation methods enabled wide involvement and innovation. In Belfast, by contrast, processes were quality-oriented and deterministic, with detailed procedures and a standardised and traditional approach to consultation, but this arguably conferred objectivity and fairness in the context of fractured communities and interests. Consequently, differences in procedures and mechanisms are explained in part by the different context and institutional sponsors.

In terms of the role and views of artists, in the Manchester case, many expressed a preference for more detailed guidance, since some found the flexibility of the process confusing. Moreover, in the Belfast case, the artists appeared to appreciate detailed guidance and procedures, and considered the selection process to be fair (Laganside Corporation, 2002). This indicates that firm and detailed frameworks and procedures are broadly desirable since they set clear parameters. This supports the suggestions of Wansborough and Mageean (2000) for more explicit design frameworks and strategies for public art. This could potentially assist with ensuring that public art meets all its aims more effectively, including those related to tourism attractions.

This however raises the issue of the role of the institutional sponsor, with their control over the form and content of guidance. There may be tensions concerning the priority to be given to local identity as opposed to external image, and this may be influenced by the sponsor. There may also be differing priorities for involvement of local communities and stakeholders, linked to institutional sponsorship. There would also seem to be scope for involving artists more in brief construction, albeit within the context of a wider strategic framework for public art, again linked to aims for tourism and place promotion.

Conclusions

While 'culture-led' regeneration as applied via cultural quarters or clusters, including aims for place promotion, have become embedded in the orthodoxy of regeneration policy, aspects of such approaches are contested. Public art provision in cultural quarters, and its linkage with local identities, is perhaps an example of this, particularly in terms of potential conflict with aims for image enhancement (potentially linked to attraction of investment and development, but also to attraction of tourists and visitors) and local identity (potentially linked to cultural tourism in particular). Such public art provision also involves some conflicting assumptions such as flexible processes versus detailed guidance.

The cases above consider these issues, and they suggest that objectives reflecting local identities and place promotion can be effectively integrated, though the balance between each is a function of different circumstances. They also show that public art can contribute to 'culture-led' regeneration by encouraging a distinctive creative milieu (particularly in areas of cultural production) and tourism (particularly 'cultural tourism'), as well as enhancing quality of life. They also show that the use of historical associations can link public art with local identity and context, but they suggest the need for more detailed guidance and strategies/frameworks for public art, and more involvement of local communities, if public art is to align more effectively with economic, social and environmental aims. This should allow the use of public art within cultural quarters to achieve broad regeneration aims, and meet the needs of a wide variety of communities and stakeholders, including tourists and visitors.

Acknowledgements

The contribution of the University of Dundee Faculty of Arts and Social Science Small Grants Scheme to the funding of the research project from which this chapter is derived is gratefully acknowledged. Primary data was also reported in McCarthy, J. (2006b), below.

References

Ashworth, G. and Voogt, H. (1990). *Selling the City*. London: Belhaven Press.

Avraham, E. (2004). Media Strategies for Improving an Unfavourable City Image. *Cities*, 21(6), 471–479.

Belfast City Council (2004). *Belfast Metropolitan Area Plan*. Belfast: Belfast City Council.

Bell, D. and Jayne, M. (2004). *City of Quarters: Urban Villages in the Contemporary City* Aldershot: Ashgate.

Breen, A. and Rigby, D. (1996). *The New Waterfront: A Worldwide Urban Success Story*. London: Thames and Hudson.

Briggs, S. (2005). *How You can Benefit from Cultural Tourism. Visit Britain Advisory Guide*. London: Visit Britain.

Bristol City Council (2000). *The Bristol Public Art Strategy*. Bristol: Bristol City Council.

Brown, E. (2008). Waterfalls Made Cash Flow? City Says Art Project Brought $69 M. *New York Observer*. Retreived 30 September 2012 from <http://observer.com/2008/2010/waterfalls-made-cash-flow-city-says-art-project-brought-2069-m/>.

Cameron, S. and Coaffee, J. (2005). Art, Gentrification and Regeneration – From Artist as Pioneer to Public Arts. *European Journal of Housing Policy*, 5(1), 39–58.

Department of Culture Media and Sport (1999). *Guidance for Local Authorities on Local Cultural Strategies*. London: Department of Culture, Media and Sport.

Evans, G. (2003). Hard-Branding the Cultural City – from Prado to Prada. *International Journal of Urban and Regional Research*, 27(2), 417–441.

Evans, G. (2004). Cultural Industry Quarters: From Pre-Industrial to Post-Industrial Production, in D. Bell and M. Jayne (eds), *City of Quarters* (pp. 71–92). Aldershot: Ashgate.

Griffiths, R. (1999). Artists' Organisations the Recycling of Urban Space, in L. Nystrom (ed.), *City and Culture: Cultural Processes and Urban Sustainability* (pp. 460–475). Karlscrona: Swedish Urban Environment Council.

Hall, T. and Robertson, I. (2001). Public Art and Urban Regeneration: Advocacy, Claims and Critical Debates. *Landscape Research*, 26(1), 5–26.

Hall, T. and Smith, C. (2005). Public Art in the City: Meanings, Values, Attitudes and Roles, in M. Miles and T. Hall (eds), *Interventions: Advances in Art and Urban Futures* (vol. 4, pp. 175–179). Bristol: Intellect Books.

Hamilton, J., Forsyth, L., and de Iongh, D. (2001). Public Art: A Local Authority Perspective. *Journal of Urban Design*, 6(3), 283–296.

Julier, G. (2005). Urban Designscapes and the Production of Aesthetic Content. *Urban Studies*, 42(5/6), 869–887.

Laganside Corporation (2000). *Cathedral Quarter Art in the Streetscape: Art Brief*. Belfast: Laganside Corporation.

Laganside Corporation (2001). *Laganside Public Art Strategy*. Belfast: Laganside Corporation.

Laganside Corporation (2002). *Public Art Policy: Equality Impact Assessment Report*. Belfast: Laganside Corporation.

London Borough of Croydon (2005). *Supplementary Planning Guidance*. London: London Borough of Croydon.

Matarasso, F. (1997). *Use or Ornament? The Social Impact of Participation in the Arts*. Stroud: Comedia.

McCarthy, J. (1998). Dublin's Temple Bar: A Case Study of Culture-Led Regeneration. *European Planning Studies*, 6(3), 271–281.

McCarthy, J. (2006a). Promoting Image and Identity in Cultural Quarters: The Case of Dundee. *Local Economy*, 20(3), 280–293.

McCarthy, J. (2006b). Regeneration of Cultural Quarters: Public Art for Place Image or Place Identity? *Journal of Urban Design*, 11(2), 243–262.

Miles, M. (1997). *Art, Space and the City: Public Art and Urban Futures*. London: Routledge.

Miles, M. (2005a). Interruptions: Testing the Rhetoric of Culturally Led Urban Development. *Urban Studies*, 42(5/6), 889–911.

Miles, S. (2005b). 'Our Tyne': Iconic Regeneration and the Revitalisation of Identity in NewcastleGateshead. *Urban Studies*, 42(5/6), 913–926.

Mommaas, H. (2004). Cultural Clusters and the Post-Industrial City: Towards the Remapping of Urban Cultural Policy. *Urban Studies*, 41(3), 507–532.

Montgomery, J. (1995). The Story of Temple Bar: Creating Dublin's Cultural Quarter. *Planning Practice and Research*, 10, 101–110.

Montgomery, J. (2003). Cultural Quarters as Mechanisms for Urban Regeneration. Part 1: Conceptualising Cultural Quarters. *Planning, Practice and Research*, 18(4), 293–306.

Neill, J. (2004). *Urban Planning and Cultural Identity*. London: Routledge.

Nowakowski, O. and Walker, R. (1999). *Applied Art in the Northern Quarter: An Evaluation*. Unpublished report.

Porch, R. (2000). Public Art – An Off the Wall Proposition? *Urban Design*, 76, 16–19.

Richards, G. (2001a). The Development of Cultural Tourism in Europe, in G. Richards (ed.), *Cultural Attractions and European Tourism* (pp. 3–30). London: CABI.

Richards, G. (2001b). The Experience Industry and the Creation of Attractions, in G. Richards (ed.), *Cultural Attractions and European Tourism* (pp. 55–70). London: CABI.

Scott, A. J. (2004). Cultural-Products Industries and Urban Economic Development: Prospects for Growth and Market Contestation in a Global Context. *Urban Affairs Review*, 39(4), 461–490.

Scottish Executive (2000). *Scotland's National Cultural Strategy*. Edinburgh: Scottish Executive.

Selwood, S. (ed.) (1995). *The Benefits of Public Art*. London: Policy Studies Institute.

Sharp, J., Pollock, V., and Paddison, R. (2005). Just Art for a Just City: Public Art and Social Inclusion in Urban Regeneration. *Urban Studies*, 42(5/6), 1001–1023.

Wansborough, M. and Mageean, A. (2000). The Role of Urban Design in Cultural Regeneration. *Journal of Urban Design*, 5(2), 181–197.

Williams, C. (1997). *Consumer Services and Economic Development*. London: Routledge.

Zukin, S. (1995). *The Cultures of Cities*. London: Blackwell.

Chapter 15
Branding Montmartre: Tourism, Popular Culture, and the Artistic Bohemia

Alexander Vari

Introduction

For visitors to Paris, the neighbourhood of Montmartre, also known because of its higher elevation within the city as the *Butte* (mound, hillock), conjures up the images of the *Cabaret du Chat Noir [Black Cat Cabaret]*, of the French *Cancan* at the *Moulin Rouge*, of bohemian painters with their canvasses placed at street corners, and, more generally, of a Parisian *fin-de-siècle* rich in cultural developments. Montmartre of yesteryear is no more but its memory lives on through the debris of its past scattered in the four corners of the world. The *Cabaret du Chat Noir* has closed its doors long ago but images of the black cat drawn by Théophile Alexandre Steinlen and of the Montmartre revellers painted by Adolphe-Léon Willette that decorated the original establishment are to be found as posters that one can purchase today either in Montmartre, or in Moscow, Tokyo, Buenos Aires, on American college campuses, or on the Internet. The figure of Aristide Bruant, a popular French singer, who was once the attraction at Le Mirliton [Reed Pipe], another Montmartre cabaret, is also a frequently re-occurring image on these posters. Although the original *Moulin Rouge* was demolished at the beginning of the twentieth century, the building of the second *Moulin Rouge* that replaced it in 1903 still stands and functions as a revue theatre. However, it is not the *Moulin Rouge* in its contemporary incarnation that stands as a symbol for Montmartre, but rather, the original's successive recreations in Hollywood films. The work of Vincent Van Gogh, Henri de Toulouse-Lautrec, Pablo Picasso and Amedeo Modigliani, who once lived and painted in the neighbourhood, is now displayed in the most important museums in the world and can be found in Montmartre only in the form of cheap reproductions that tourists buy in Place du Tertre, a square near the Basilica of the Sacré Coeur, on top of the *Butte*, where the latter congregate in large numbers.

In order to promote tourism, today's Montmartre feeds off of memories of its golden age, but is also, as these examples demonstrate, a place that is constituted through a powerful set of global interactions and discourses. Montmartre is therefore not just a physical space shaped by nostalgia for its past but a cultural construct, one that was constituted through the addition of several layers of meaning to what *fin-de-siècle* Montmartre originally meant, within the framework of a continuously evolving historical context shaped by the actions, desires and discourses of both locals and visitors. With these important parameters in mind, this chapter argues more specifically that Montmartre's turning into a Parisian and global cultural symbol was also the outcome of a long-term place branding process. Motivated by a desire to promote the neighbourhood as a touristic destination, place branding was an endeavour that became prominent as early as the 1880s. Participants in it were the cultural actors themselves, whose place was later taken by memoirists, film makers, and media conglomerates, who, as I discuss later in this chapter, kept the myth of *fin-de-siècle* Montmartre alive for worldwide audiences.

Local Cultural Effervescence and the Making of Montmartre Brands

The artistic cabarets of *fin-de-siècle* Montmartre represented an important chapter in the history of modern Western art. In them 'popular music, joined forces with art and literature in a synthesis of high and low cultures that has since rarely been equaled' (Gendron, 2002, p. 29). The songs, performances and artistic work produced in cabarets such as the *Chat Noir*, the *Mirliton*, the *Divan Japonais [Japanese Lounge]*, the *Cabaret des Quatz'Arts [Cabaret of the Four Arts]*, including several others (see Oberthür, 1984), constituted, according to Gendron (2002) '"secondary" aesthetic practices' which, as a result of 'sharing the same intertextual space, participated quite significantly in determining the public reception of the "primary" modernist products [such as paintings, poems, and symphonies], and hence in the constitution of their public meanings' (p. 21). The cabarets and the poets, song writers, painters, and singers who animated them called to life what became known as the spirit of Montmartre: an eclectic mix of satire, humour, and performance which anticipated later trends in modern avant-garde art (see Rearick, 1985; Seigel, 1986; Cate, 1999; Weisberg, 2001; Franck, 2003).

In *fin-de-siècle* Montmartre artistic pursuits mixed with commercial ones (see Kenny, 2004). The artistic cabarets were the first establishments that put Montmartre on the tourist map. The *Chat Noir*, the first of these to open in December 1881, came into being as a result of the move of a group of young artists and literati from the Latin Quarter to Montmartre, a neighbourhood that up to then was marginal within Paris's urban and cultural topography (Goudeau, 1888). Its marginality, however, could be best understood if contrasted with the centrality of the Great Boulevards. The latter were the area where most of the city life was concentrated and where the largest crowds of locals, provincials and foreign visitors congregated to stroll, eat in cafes and restaurants, and watch operatic productions and theatrical plays (see Hahn, 2009). Therefore, the *Chat Noir* took up, as voiced by its artist-cum-businessman owner, Rodolphe Salis, the ambivalent goal of rejecting the consumption-centred popular culture of the boulevards, while recreating it in a different guise in the midst of a still non-commercial and non-touristy neighbourhood. Everything that the *Chat Noir* did was based on this duality. While the tourism and novelties-oriented culture of the boulevards was openly mocked in the *Chat Noir* and in the pages of the eponymous journal which Salis published, in an attempt to appear different, the cabaret appealed to its visitors through an exotic interior based on the mixing of fake medieval furnishings with pieces of art produced by Salis's collaborators (for the gothic revival promoted by the *Chat Noir* see Debans, 1889; Grand Carteret, 1886; and Emery and Morowitz, 2003).

Members of the artistic bohemia who congregated around Salis were fond of the village-like character of the neighbourhood in which the *Chat Noir* was located. The emphasis on Montmartre's rural aspect served to further distance it from Paris. But they were also proud of their role as purveyors of a new type of urban popular culture and entertainment (see Gluck, 2005), which as a result made the neighbourhood to be marketed by them as a place with rural aspect but highly urban manifestations. For instance, in one of the first issues of the *Chat Noir*, the journal that Salis published as a mouthpiece of his cabaret, one of its collaborators stated that 'Montmartre is the cradle of humanity … Montmartre is the center of the world … It was in Montmartre that humans built the first city' (Le Chat Noir, 1882, January 14). The cultural effervescence in the world of art, song and poetry generated by the *Chat Noir* together with the bombast of statements such as these, caught the attention of Parisians. The jocular and playful spirit of the *Chat Noir* created a veritable vogue among members of the Parisian high life prompting many well-known writers, artists and politicians to make it to the cabaret, to be subtly mocked there by the proprietor (Bloy, 1884).

While interest in Salis's establishment grew continuously during the 1880s, it was the universal exhibition held in Paris in 1889 that opened the floodgates of mass tourism. Many of the Paris guides published that year, either at home or abroad, mentioned the *Chat Noir* as a must see for both French and foreign tourists (*Paris sa vie et ses plaisirs par un Parisien du Pré aux Clercs*, 1889). Tourist guides focusing on the French capital's night-time attractions were even more detailed, including long chapters about it. For instance, in his *Nights in Paris*, Rodolphe Darzens (1889) presented the *Chat Noir* as a cabaret that had to be visited in order for one not just to enjoy a pleasurable evening about town listening to the songs and comic skits or watching the shadow plays showed there, but as a veritable museum where one could also see the famous paintings by Willette and Steinlen. Indeed, fulfilling the wishes of his founder, Salis's establishment turned into 'a journal, an exhibition, a concert hall, a theater, [and] a coffeehouse', all in one, a museum-like place of pilgrimage for everyone in Paris (Debans, 1889, p. 145).

However, by 1889 the *Chat Noir* was just one among more than a dozen cabarets where one could spend time in Montmartre. After Salis moved his cabaret in 1885 to a new quarter just a few blocks away, the locale of the original establishment on Boulevard du Rochechouart was taken over by Aristide Bruant, a pianist and singer who often played at the *Chat Noir*, and who reopened it under a new name, *Le Mirliton*. Unlike the *Chat Noir* which continued as a group enterprise headed by Salis covering a broad artistic range and having a museum-like character, the Mirliton became Bruant's self-aggrandising temple and stage where the attraction was the singer himself. Wearing a larger than life hat and red scarf wrapped around his shoulders and dressed every night in a patent red shirt, black ribbed velvet hunting coat, pants and leather boots, Bruant's songs in French argot translated the sufferings of the neighbourhood's homeless and downtrodden to a bourgeois audience, whom he constantly mocked and verbally abused (Valbel, 1895; Wilson, 2008).

Other literary and artistic cabarets which opened during these years, such as the *Divan Japonais* and the *Cabaret de Quat'z Arts*, tried to live off the success of the *Chat Noir*. However, while the artistic cabarets of Montmartre evolved into a neighbourhood brand, which captured a lot of public attention throughout the 1880s, during the next decade interest shifted towards the neighbourhood's dance halls. The opening in October 1889 by Joseph Oller, a Catalan businessman naturalised in France and Charles Zidler, his artistic director and associate, of the *Moulin Rouge* on Place Blanche, in the immediate vicinity of the cabarets mentioned above, brought to the fore a new type of entertainment. The French *quadrille* (soon to be known in the English-speaking world as the French can-can) became an instant sensation, which stole much of the appeal of the cabarets. The *Moulin Rouge's* dancers wearing popular nicknames such as La Goulue [The Glutton], Nini Patte en l'air [Nini Foot-in-the-Air], Jane Avril, Grille d'Égout [Drainage Grating], La Môme Fromage [The Cheese Kid], La Sauterelle [The Grasshopper], Rayon d'Or [Ray of Gold], including a male dancer, Valentin le Désossé [Valentin the Boneless], rose rapidly as stars of the establishment, attracting in the early evening a lower-class clientele made up of local residents stretching from 'menial workers, shop-assistants, [and] dressmakers' to the 'residents of Place Blanche' to be followed as the evening progressed by 'the arrival of … tradespeople and businessmen seeking to forget their worries, students, elegant women, lords and ladies, ambassadors, royalty, visiting foreigners and those whose social standing obliged them to adopt a subterfuge for frequenting low company' (Pessis and Crépineau, 1990, p. 33).

But what turned the *Moulin Rouge* into a place competing with the cabarets in the branding of Montmartre as a neighbourhood of nocturnal pleasures and entertainment was this dance hall's prominent featuring in the works of Henri de Toulouse-Lautrec. Of ancient aristocratic pedigree but suffering of numerous physical deformities, after moving from south-western France to start an artistic career in Montmartre during the 1880s, Toulouse-Lautrec became fascinated with *Moulin Rouge*'s dancers whom he immortalised in several posters and paintings he produced during the 1890s. Many of them, such as his famous 1891 modernist poster depicting La Goulue and Valentin le Désossé, were displayed in thousands of copies on Paris's Morris columns and its mural advertising spaces, contributing to the *Moulin Rouge*'s draw of visitors from all corners of France and the world (Thompson, Cate, Weaver Chapin, and Coman, 2006). During these years, the red wings of the fake windmill in which Oller's and Ziedler's dance hall was located turned into a Montmartre icon that competed with the Eiffel Tower in defining the meaning of Parisianness for those living abroad.

Although Montmartre was a place where there were numerous other literary cabarets and dance halls, it was the *Chat Noir*, the *Mirliton* and the *Moulin Rouge* that turned into a distinct neighbourhood brand. It is because of this, that as a historian of Montmartre's rich cultural heritage writes 'the posters of Le Chat Noir, Aristide Bruant, and Jane Avril stand out today as visual markers of the butte, and each tourist who returns from Montmartre with reproductions of famous works believes he has preserved for himself the essence of the pilgrimage through the image' (Weisberg, 2001, p. 1).

The Commercialisation of Montmartre and the Making of New Neighbourhood Brands

By the early 1900s, however, it was not anymore the artistic quality and originality but rather the shock value and the commercial character of many of the cabarets' offerings that fuelled tourism to the *Butte*. After Bruant's desertion from the Mirliton in 1895 and the closing down of the *Chat Noir* two years later, the shift in this direction became obvious. Newer cabarets such as those of *Ciel [Heaven]* and *Enfer [Hell]* thrilled visitors with interiors and shows which took the by then traditional display of 'spectacular realities', inaugurated in Paris by wax museums such as the *Musée Grévin* and the spectacle of the Parisian morgue, to a new level of public titillation (see Schwartz, 1998). Located next to each other, the cabarets of *Ciel* and *Enfer*, played on their physical proximity to offer visitors a quick experience of heaven and hell. Their exteriors contrasted greatly from one another. While the *Ciel*'s façade was painted in white and blue and had angels decorating it, that of the *Enfer* was red and black and was dominated by the huge plaster mouth of a Satan-like figure. Replicating this thematic emphasis, the waiters of *Ciel* were dressed either as saints or 'white-robed angels with gauze wings and brass halos' while in the establishment next door they served customers in the guise of demons and devils (see The fantastic cafés of Paris, 1902, p. 594).

Another new addition to the attractions of the early twentieth-century *Butte*, *Le Cabaret du Néant [The Cabaret of Nothingness]*, was even more sensationalist. According to one account by a visitor in 1899:

> As you enter, you find yourself in a room shaped and fashioned like a mortuary hall. In place of tables are coffins on trestles, lighted by tiny corpse candles. Men in the garb of undertakers greet you, and tell you to choose your seats, and order your drink, commenting on the quality of their beer and wine in a manner revolting to hear and impossible to describe. In place of a chandelier, there is an arrangement of human bones; a human skeleton, with battered hat on head, and short clay pipe between its grinning jaws, mocks you from one wall. Skulls hang from the ceiling, and there are strange pictures all round. (Bensusan, 1899, p. 235)

Although even after its demise in 1897, the *Chat Noir* was still celebrated as a model for and ancestor of all the Montmartre cabarets, the crowds of early twentieth-century American tourists arriving to Europe and Paris by steamboat were more likely to form their image of the Montmartre bohemia and artistic underworld based on their experiences with the *Ciel*, *Enfer* and *Néant* cabarets (Anon., 1899). According to some visitors, however, the displays that one could watch in these establishments were 'cold, tawdry and very stupid' (Lucas, 1909, p. 270). For many American tourists they evoked the sensationalist and commercial offerings of Coney Island at home (Wells, 1907). More sophisticated foreigners were aware that they were just tourist-traps, which French people largely avoided. What they were looking for were the ghosts of the past: those of the *Chat Noir* during its golden age (Mencken, Nathan, and Huntington Wright, 1914).

One response to the closing of the *Chat Noir* and the commercialism that engulfed the newer cabarets was indeed this growing nostalgia and longing among visitors for the fast receding past of the *Butte*, while another one came from elsewhere. In addition to its cabarets and dance halls, the *Butte* also housed an important art colony. Already during the late 1880s and 1890s Montmartre-themed landscapes and streetscapes proliferated as artists set their canvasses on street corners to capture the sights of the *Butte*. Among them there were artists well known today such as Vincent Van Gogh and Camille Pissarro and lesser known ones as the American impressionist painter Childe Hassam. These painters dissociated themselves from other Montmartre artists (such as Willette, Steinlen, Toulouse-Lautrec, Louis Morin, Caran d'Ache, André Forain) who were preoccupied to decorate the interior of the cabarets, record the goings on in the dance halls, portray the picturesque characters of the neighbourhood and publish their work in the numerous literary and art journals that were published in Montmartre (see Emanuel, 1904).

The discovery of Montmartre by impressionist and post-impressionist painters, however, went quite unnoticed as most of the attention of visitors focused on the cabarets and the dance halls. The international vogue of the latter, attracted not just tourists but hundreds of foreign artists (with Picasso, Kees van Dongen, Modigliani, Juan Gris, and others, soon to become the most well-known among them). The arrival of these artists turned the neighbourhood into an important headquarter of avant-garde art. Many of them rented studios in the ramshackle but cheap Bateau-Lavoir on Place Ravignan (Warnod, 1975) and chose to congregate high up on the hill in the bistros of the Place du Tertre and in the *Cabaret du Lapin Agile [The Nimble Rabbit Cabaret]*, a place off the beaten path in its vicinity. All these were places where a new bohemia also came into being during these years. Its members included not just Picasso, Van Dongen, and Modigliani but French artists and writers such as Guillaume Apollinaire, Max Jacob, Maurice Utrillo, André Salmon, Pierre Mac Orlan, Francis Carco, Roland Dorgelès, and André Warnod, most of who were born during the 1880s (see Seigel, 1986; Lay, 2001).

Although many of Picasso's canvasses painted during his stays in Montmartre or at the Bateau-Lavoir (such as *La Vie [Life]*, *Les Demoiselles d'Avignon [The*

Young Ladies of Avignon]) are justly celebrated as master works (Richardson, 1991), it was not these paintings that turned into a new Montmartre brand but those of Maurice Utrillo. Born as the son of Suzanne Valadon, another well-known Montmartre-based painter, Utrillo was the quintessential bohemian. Taken to drink from an early age, he was often put under restraint by his mother (Coquiot, 1925). Utrillo's subject when he painted were the narrow and winding Montmartre streets going uphill to the Place du Tertre, the picturesque character of which was slowly disappearing as the face of the neighbourhood was changing under the impact of new buildings erected on its slopes by real estate speculators. Alike to the *Chat Noir's* literary soirées and Bruant's songs more than two decades before, which were instrumental in the rise of what became known as the Montmartre spirit, and Toulouse-Lautrec's 1890s posters that advertised the *Moulin Rouge*, Utrillo's paintings were important in drawing fresh attention to the neighbourhood. However, the way the *Butte* was presented this time was different. Instead of the poetry, dance and song that became the staples of late nineteenth-century Montmartre the emphasis fell on the neighbourhood's physicality which was turned into a new neighbourhood brand through the medium of painting. Utrillo's success emboldened dozens of less talented artists to paint in the same style and manner, a practice that made Montmartre streetscapes ubiquitous in the restaurants, bistros and art galleries of the time.

Utrillo's companions at the *Lapin Agile* also played an important role in devising a new Montmartre brand. Throughout the interwar period and even as late as the 1950s and 1960s, with their portraits of avant-garde artists and prolific recollections of their life in the neighbourhood before 1914, Carco, Mac Orlan, Warnod, and Dorgelès brought to life a literary myth of Montmartre bohemia. For them real Montmartre was neither the *Chat Noir* of the 1880s and the *Moulin Rouge* of the 1890s, nor the world of the Americanised music halls and cabarets where jazz resounded and tourists swarmed throughout the interwar period, but the young spirit of the writers and poets who frequented the *Lapin Agile*. In their view, urban development, Americanisation, commercialisation and mass tourism, which started in the first years of the century and completely overtook the neighbourhood by the interwar period (when cabarets and dance halls were replaced by film palaces), were the ills of Montmartre that one could only fight by flying back on the wings of nostalgic recollections to the exciting time that they had lived as twenty year-olds on the upper slopes of the *Butte* (Vari, 2008). The conservative overtones of this neighbourhood portrayal are hard to miss. It is not by chance that the Anglo-Saxon and Afro-American experience of Montmartre during the roaring twenties celebrated in the writings of American expatriates or Afro-American jazz musicians visiting or living on the *Butte* during that period (McMahon, 1964; Shack, 2001) is largely ignored by these French writers, and if featured, presented under the guise of destructive forces that were corrupting the face and spirit of the neighbourhood.

As Carco, Mac Orlan, Warnod, and Dorgelès's nostalgic portrayal of pre-war Montmartre grew into a new neighbourhood brand, the *Lapin Agile* and the

Place du Tertre became more and more popular with tourists. The popularity of Utrillo's paintings and that of hundreds of his imitators made both places easy to identify and triggered a desire among many to visit them. The anti-touristic perspective of those who rejected low Montmartre in the early twentieth-century became something that visitors to the *Butte* celebrated by crowding into the bohemian haunts of the second-generation of Montmartre artists. As one of the latter's members wrote, the first consistent forays into the *Lapin Agile* were made by American doughboys during the late phase of WWI making it to resemble 'a bar from the Far West' (Warnod, 1920 in Blower, 2011, p. 66). Although during the mid-1920s American visitors to Montmartre made it to the Lapin Agile and the adjacent Place du Tertre mostly by day, and for most of them the *Butte* was still 'a nighttime destination centered on a cluster of clubs at the base of the hill' (Blower, 2011, p. 44), and Montparnasse, a neighbourhood at the opposite corner of Paris, where many members of a cosmopolitan bohemia congregated both before and after the Great War, represented a new pole of attraction for them, their wholesale takeover of the *Butte* was completed during the next decades.

Pre-war Montmartre Nostalgia as Neighbourhood Brand in Post-war Popular Culture

The Allied liberation of Paris on August 23, 1944 removed the temporary obstacle that the German occupation of the French capital four years earlier and the raging battles of WWII created in the way of tourists wishing to visit Montmartre from countries not aligned with the Axis Powers. The presence of American GIs on the *Butte* just a few hours after the liberation of the French capital bode well for Montmartre cafes, bars, brothels and other businesses (Chevalier, 1985). While during the war, the Champs-Elysées took away from Montmartre its role as centre of Paris's nocturnal life, after the war the neighbourhood turned again into a place visited by many tourists. This time, however no new bohemia emerged to add further attractions to the offerings of the *Butte*. Instead of turn-of-the-century Montmartre and interwar Montparnasse, Paris literati and bohemians met now in the fashionable cafes and jazz dives of the Saint-Germain des Près neighbourhood on Paris's Left Bank. As a result of this new shift in Paris's cultural topography (see Hewitt, 1996), in Montmartre the emphasis fell again on the appeal of old haunts, memories, images and representations.

During the early post-war period, Hollywood came to play a leading role in the fostering of this new wave of nostalgia. A spate of films with Parisian themes was produced by American film companies during the 1950s. A couple of them, such as *Moulin Rouge* (1952, dir. John Huston) and *Can-Can* (1960, dir. Walter Lang) were shot on location in Montmartre. *Moulin Rouge* had a dance and song-rich plot that made viewers believe that, as the 'film's scrolling narrative explain[ed] "for a brief moment … Toulouse Lautrec, his beloved city and time live again"' (Schwartz, 2007, p. 23). In *Can-Can,* Frank Sinatra and Maurice Chevalier sang 'about the joys of "Montmartre" in 1890s Paris' (Santopietro, 2008, p. 274). Together with other films like *An American in Paris* (1951, dir. Vincente Minelli), which also included snapshots of Montmartre and the Moulin Rouge, these filmic recreations of late nineteenth-century life on the *Butte* were watched by millions of spectators in film theatres worldwide, visually preparing those among them who would go to Paris for what to look for once there.

Many of these American musicals came on the heels of French films shot in Montmartre and elsewhere during the 1930s, and were competing with such French post-war features as Jean Renoir's *French Can-Can* (1954). The invasion of the *Butte* by seekers of nocturnal pleasures in the immediate aftermath of WWI, the disappearance of old Montmartre under the hammer of modern real estate developers together with the replacement of former cabarets and dance halls with film palaces during the late 1920s (Chevalier, 1980; Jackson, 2006; Vari, 2008), developments which, as we have seen, radically reshaped the physical geography of pre-war Montmartre, motivated many French film directors during the 1930s to try and recapture its *fin-de-siècle* spirit through a strong featuring of realist popular songs in their films. Unlike in the Hollywood films of the 1950s and early 1960s, in which, as Vanessa Schwartz (2007) argues, elements of Frenchness were integrated and became part of a transnational film culture, in these films Montmartre was turned into an instrument for the preservation of a French identity seen as being under attack by the spread of an America-driven process of cultural globalization. A number of famous French realist singers of the interwar period such as Fréhel, Odette Barencey and Damia, cast in films like *Pépé le Moko* (1936, dir. Julien Duvivier), *Faubourg Montmartre* (1931, dir. Raymond Bernard) and *La Tête d'un homme [Head of a Man]* (1932, dir. J. Duvivier), came to stand for the intimately French character of the neighbourhood through their songs sang in Parisian argot which continued Bruant's musical tradition. They became 'symbol[s] of nostalgia in the 1930s for the café-concert, and for the pre-World War I era more generally' (Conway, 2004, p. 100).

With their nostalgic portrayal of the *Belle Époque*, the French films of the 1930s catered thus to a French audience that was mourning the loss of French traditions. Their nostalgia for a lost Montmartre was also powerfully re-enacted in Renoir's *French Can-Can*, filmed in a Montmartre location two decades later. However, this time the context was different. Instead of focusing on the French realist musical tradition and the preservation of Frenchness, Renoir's film made a strong bow to Hollywood with its 'exaggerated, artificially "picturesque" quality that suggest[ed] a Disneyland re-creation' of turn-of-the-century Montmartre. 'At times – as film critic Dave Kehr writes – Renoir seems to be peddling a tourist's fantasy of gay Paree' (Kehr, 2011, p. 269). While the majority of film critics during the 1950s celebrated the Hollywood films about Paris and Montmartre, a few pointed out how by focusing on an idealised past they ignored the problems of a present, and especially in the case of Jean Renoir's *French Can-Can*, the latter's carefree portrayal of turn-of-the-century life in Montmartre enabled the French to forget about their immediate collaborationist past during the Vichy regime in WWII (Bergstrom, 1998), as well as about the new reality that the new post-war avant-garde was not located anymore in Paris's Montmartre or Montparnasse but New York's Manhattan (on the latter see Guibaut, 1985). Last but not least, the film's plot and its focus on the Moulin Rouge, with its fictional character Danglard (modelled on the original dance hall's founder Joseph Oller) intent on making money, as another analyst of the film pointed out 'showed the beginning of the contemporary Parisian "package tour" industry … and that France needed foreigners to function economically – at the beginning for investment capital and later as clients' (Bergstrom, 1998, p. 211).

Renoir's film, together with the Paris films produced by American studios during the 1950s and early 1960s, stood indeed for the new transnational character of the global film industry (see Schwartz, 2007). In them Montmartre became a Hollywood trope which was taken up in later films about Montmartre as the *Butte* continued to decay and loose its local colour during the last decades of the twentieth-century. The globalisation of the neighbourhood, which with its sex shops, traffic jams, T-shirt shops and con artists posing as Montmartre bohemians came to resemble any other sex district in 1970s and 1980s New York, London and Hamburg (see Chevalier, 1985), was part of a process that was replicated in the reverse direction by the global film industry for which representations of the *Butte* depended on a Disneyesque recreation of local colour. While Montmartre lost its touristic appeal on the ground, its lure was resurrected in films, with a virtual reality marketed as a Hollywood brand superseding the reality of the neighbourhood.

This development was especially noticeable in two films released in 2001: the international production *Moulin Rouge* (dir. Baz Luhrmann) and the French *Le Fabuleux destin d'Amélie Poulain* (dir. Jean-Pierre Jeunet), with the latter marketed in the United States as *Amélie from Montmartre.* They continued the nostalgic tradition of Hollywood films about Montmartre but this time with the use of much more sophisticated digital technologies. As the author of a recent book about Baz Luhrmann's work writes, in Luhrmann's *Moulin Rouge* 'three distinct periods characterised by social, cultural and technological upheaval (fin-de-siècle Paris, 1950s Hollywood and the modern global era) are synchronised to create an imaginary world where temporal and spatial boundaries are dissolved and obstacles can be overcome' (Cook, 2010, p. 101). Filmed in Twentieth-Century Fox's studios in Sydney, Australia, Luhrmann's film is a flashy and flamboyant recreation of the world of *fin-de-siècle Moulin Rouge* in a dazzling combination of Hollywood's and Bollywood's cinematic language and MTV's video clip styles. Instead of receding into the past, turn-of-the-century Montmartre becomes unbound in space and timeless.

Although shot on location and using a different approach and techniques, Jeunet's *Amélie* achieves the same effect. As IsabelleVanderschelden, the author of the first scholarly monograph about the film, points out:

> *Amélie* mythologizes Montmartre at least as much as its captures its reality, resorting to elusive truth, selective fragmentation and cosmetic changes. In many exterior shots, the modern urban elements that do not fit in with the

> fairytale atmosphere of the film have been deleted. The result is an idealized representation of Montmartre as a clean non-threatening version of Paris, a playground with funfairs and merry-go-rounds, championing a convivial village-like atmosphere. (Vanderschelden, 2007, p. 69)

Known mostly as an independent filmmaker before the making of *Amélie*, and having lived in Montmartre since the 1970s, Jeunet's representation of the neighbourhood 'reflects his own ambivalent perception of the place, as the picturesque and magical part of [Paris] … seen by the outsider that he once was on the one hand, and the centre of a village-like community that he has now become integrated with on the other' (Vanderschelden, 2007, p. 67). Though accessing Montmartre from a different vantage point (emphasis on quirky French characters as an attempt to capture local colour and a love plot centred in Montmartre's present, instead of Luhrmann's recycling of the *fin-de-siècle* atmosphere), Jeunet's film relies on a transnational filmic language to rescue the neighbourhood from oblivion. Even more to the point, *Amélie* can be read as another rejection of a disturbing reality (a Montmartre disfigured by sex shops, graffiti, modern buildings, mass commerce and tourism), including a conscious attempt to revive the neighbourhood's tourist economy by making it appealing again for foreign visitors.

Many commentators agree that these films are furthering earlier processes of Paris (and within it Montmartre's) Disneyfication for tourist consumption, culminating in the recent recreation of Paris as a Las Vegas theme park. However, nostalgia for a timeless Montmartre can be caught at work not just in the recreation of Paris in the Nevadan dessert, but in the fostering of a strong desire among many global tourists to see the original. Building on Susan Stewart's work on longing and Guy Debord's take on the society of spectacle, Katherine Lawrie van de Ven has recently argued that nostalgia as 'desire for desire' and 'the form of desire most amenable to spectacle … cannot be satisfied' since they 'will instead persist as melancholy in search of commodity (object or experience) fulfilment' (2010, p. 71). In this sense, a magical and timeless Montmartre, as a new neighbourhood brand marketed internationally by the global film industry, serves as continuous enticement for the consumption of the real and authentic Montmartre by hordes of contemporary tourists who crowd the streets of the *Butte* with the hope of fulfilling a desire that can never be fulfilled.

Conclusion

Since the early 1880s Montmartre has been branded under many different guises. It was a place that was successively identified with the jocular spirit of the *Chat Noir*, the argotic songs of Aristide Bruant, and the vertiginous tempos of the French Cancan performed at the *Moulin Rouge*. It has also been equated with the turn-of-the-century Parisian bohemia and celebrated as the birthplace of the modern avant-garde art. To this, Utrillo's paintings (together with those of his epigones) added a new layer that brought into focus the *Butte*'s buildings and streetscapes.

While these representations grew out as reflections of the neighbourhood's pulsing artistic life, after the war, place branding fed increasingly from its past glories. During the interwar period the branding of Montmartre became part of a nostalgic search for bygone eras which emphasised local colour and tradition against changes that occurred in the *Butte*'s present. This was equally done in the writings of former bohemians and customers of the *Lapin Agile* such as Carco, Warnod, MacOrlan and Dorgelès, and the French films of the 1930s that connected Montmartre to the argotic songs of Aristide Bruant and the world of marginal existences celebrated in them. Neighbourhood branding became thus more conservative in tone and enmeshed in sentimentalism.

After WWII the business of promoting Montmartre was picked up by film directors and media conglomerates, and place branding perpetuated in a new attempt to divert attention from the present. Similarly, more than half a century later, films like *Moulin Rouge* and *Amélie* rely on the global film and tourist market to promote Montmartre as a spectacular and magical place, thus obliterating the present-day realities of the neighbourhood, and, as some isolated voices have pointed out, disengaging both tourists and spectators from critically confronting their global present (see Kaganski, 2001). The history of the successive cultural signifiers attached to Montmartre and their turning into a neighbourhood brand, reflect thus an increasing rift between the *Butte*'s present and its highly idealised and embellished past. Shaped by such portrayals in the contemporary mass media, Montmartre is today a tourist site where its bohemian past and contribution to the making of Parisian popular culture recreated through nostalgic representations devoid of historicity continue to serve as a lure for millions of tourists.

Acknowledgments

The author would like to thank William Conlogue, Nicolas Kenny, Michael Wilson, and the book editors, for their comments and suggestions for earlier versions of this chapter.

References

Le Chat Noir (1882). [a journal].

Paris sa Vie et Ses Plaisirs par un Parisien du Pré aux Clercs (1889). Paris: Bibliothèque Chacornac.

The Fantastic Cafés of Paris (1902). In H. Quilter (ed.), *What's what: A Guide for To-day's Life as it is and Things as They are* (pp. 594). London: Sonnenschein & Co.

Anon. (1899). Chronicle and Comment. *The Bookman: A Journal of Literature and Life* (October), 212–213.

Bensusan, S. L. (1899). The World's Cafés: Paris. *The Idler, XV* (February–July), 230–236.

Bergstrom, J. (1998). Jean Renoir's Return to France, in S. Rubin Suleiman (ed.), *Exile and Creativity: Signposts, Travelers, Outsiders, Backward Glances* (pp. 180–219). Durham: Duke University Press.

Blower, B. L. (2011). *Becoming Americans in Paris: Transatlantic Politics and Culture between the World Wars*. Oxford: Oxford University Press.

Bloy, L. (1884). Le Gentilhomme Cabaretier, in *Propos d'un Entrepreneur des Démolitions* (pp. 99–105). Paris: Tresse.

Cate, P. D. (1999). *The Spirit of Montmartre: Cabarets, Humor and the Avant-Garde, 1875–1905*. New Brunswick: Rutgers University Press.

Chevalier, L. (1980). *Montmartre du Plaisir et du Crime*. Paris: Robert Laffont.

Chevalier, L. (1985). *Les Ruines de Subure. Montmartre de 1939 aux Années 1980*. Paris: Robert Laffont.

Conway, K. (2004). *The Chanteuse in the City: The Realist Singer in French Film*. Berkeley: University of California Press.

Cook, P. (2010). *Baz Luhrmann*. Houndmills: Palgrave Macmillan.

Coquiot, G. (1925). *Maurice Utrillo V*. Paris: André Delpeuch.

Darzens, R. (1889). *Nuits à Paris*. Paris: E. Dentu.

Debans, C. (1889). *Les Plaisirs et les Curiosités de Paris. Guide Humoristique et Pratique*. Paris: Ernest Kolb.

Emanuel, F. L. (1904). *The Illustrators of Montmartre*. New York: Charles Scribner's.

Emery, E. and Morowitz, L. (2003). *Consuming the Past: The Medieval Revival in Fin-de-Siècle France*. Aldershot: Ashgate.

Franck, D. (2003). *Bohemian Paris: Picasso, Modigliani, Matisse, and the Birth of Modern Art*. New York: Grove Press.

Gendron, B. (2002). *Between Montmartre and the Mudd Club: Popular Music and the Avant-Garde*. Chicago: University of Chicago Press.

Gluck, M. (2005). *Popular Bohemia: Modernism and Urban Culture in Nineteenth-Century Paris*. Cambridge, MA: Harvard University Press.

Goudeau, È. (1888). *Dix ans de Bohème*. Paris: Librairie Illustrée.

Grand Carteret, J. (1886). *Raphael et Gambrinus, ou, l'Art dans la Brasserie*. Paris: L. Westhauser.

Guibaut, S. (1985). *How New York Stole the Idea of Modern Art: Abstract Expressionism, Freedom and the Cold War*. Chicago: University of Chicago Press.

Hahn, H. H. (2009). *Scenes of Parisian Modernity: Culture and Consumption in the Nineteenth Century*. Houndmills: Palgrave Macmillan.

Hewitt, N. (1996). Shifting Cultural Centres in Twentieth-Century Paris, in M. Sheringham (ed.), *Parisian Fields* (pp. 30–45). London: Reaktion Books.

Jackson, J. (2006). Artistic Community and Urban Development in 1920s Montmartre. *French Politics, Culture & Society*, 24(2), 1–25.

Kaganski, S. (2001). '"Amélie" pas Jolie', *Libération*, 31 May. Retrieved 26 February 2013 from <http://www.liberation.fr/tribune/0101375722-amelie-pas-jolie>.

Kehr, D. (2011). *When Movies Mattered: Reviews from a Transformative Decade*. Chicago: University of Chicago Press.

Kenny, N. (2004). Je Cherche Fortune: Identity, Counterculture and Profit in fin-de-siècle Montmartre. *Urban History Review/Revue d'histoire urbaine*, 32(2), 21–32.

Lawrie van de Ven, K. (2010). Spectacular Paris: Representations of Nostalgia and Desire. *Paroles Gelées*, 26(1), 67–93.

Lay, H. (2001). Pictorial Acrobatics, in G. P. Weisberg (ed.), *Montmartre and the Making of Mass Culture* (pp. 145–179). New Brunswick: Rutgers University Press.

Lucas, E. V. (1909). *A Wanderer in Paris*. New York: Macmillan.

McMahon, J. H. (1964). City for Expatriates. *Yale French Studies*, 32 (special issue 'Paris in Literature'), 144–158.

Mencken, H. L. Nathan, G. J., and Huntington Wright, W. (1914). *Europe after 8:15*. New York: John Lane.

Oberthür, M. (1984). *Cafes and Cabarets of Montmartre*. Salt Lake City: Peregrine Smith Books.

Pessis, J. and Crépineau, J. (1990). *The Moulin Rouge*. New York: St. Martin's Press.

Rearick, C. (1985). *Pleasures of the Belle Epoque: Entertainment and Festivity in Turn-of-the-Century France*. New Haven: Yale University Press.

Richardson, J. (1991). *A Life of Picasso, volume 1: 1881–1906*. New York: Random House.

Santopietro, T. (2008). *Sinatra in Hollywood*. New York: St. Martin's Press.

Shack, W. A. (2001). *Harlem in Montmartre: A Paris Jazz Story between the Great Wars*. Berkeley: University of California Press.

Schwartz, V. (1998). *Spectacular Realities: Early Mass Culture in Fin-de-Siècle Paris*. Berkeley: University of California Press.

Schwartz, V. (2007). *It's So French! Hollywood, Paris and the Making of Cosmopolitan French Culture*. Chicago: University of Chicago Press.

Seigel, J. (1986). *Bohemian Paris: Culture, Politics and the Boundaries of Bourgeois Life, 1830–1930*. New York: Viking.

Thompson, R., Cate, P. D., Weaver Chapin, M., and Coman, F. E. (2006). *Toulouse-Lautrec and Montmartre*. Princeton: Princeton University Press.

Valbel, H. (1895). *Les Chansonniers et les Cabarets Artistiques*. Paris: E. Dentu.

Vanderschelden, I. (2007). *Amélie: Le Fabuleux Destin d'Amélie Poulain*. Urbana and Chicago: University of Illinois Press.

Vari, A. (2008). A Disappearing Montmartre: Anchoring the Memory of a Neighborhood in Interwar Paris, in J. Schachter and S. Brockmann (eds), *(Im)permanence: Cultures in/out of Time* (pp. 73–87). University Park: Penn State University Press.

Warnod, A. (1920). *Les Plaisirs de la Rue*. Paris: L'Edition Française Illustrée.
Warnod, J. (1975). *Le Bateau Lavoir, 1892–1914*. Paris: Les Presses de la Connaissance.
Weisberg, G. P. (ed.). (2001). *Montmartre and the Making of Mass Culture*. New Brunswick: Rutgers University Press.
Wells, C. (1907). *The Emily Emmins Papers ... With Illustrations by Josephine A. Meyer*: New York and London.
Wilson, M. (2008). The Eccentric Masculinity of Aristide Bruant. *The Proceedings of the Western Society for French History*, 36, 195–208.

Filmography

An American in Paris (1951, dir. Vincente Minelli)
Can-Can (1960, dir. Walter Lang)
Faubourg Montmartre (1931, dir. Raymond Bernard)
French Can-Can (1954, dir. Jean Renoir)
La Tête d'une homme (1932, dir. Julien Divivier)
Le Fabuleux destin d'Amélie Poulain (2001, dir. Jean-Pierre Jeunet)
Moulin Rouge (1952, dir. John Huston)
Moulin Rouge (2001, dir. Baz Luhrman)
Pépé le Moko (1936, dir. Julien Duvivier)

Index

Made in the USA
Monee, IL
13 September 2026

62976813R10142